Progress through Mechanical Engineering

A home of its own: Slade's designs for the new headquarters of the institution at Storey's Gate

Progress through Mechanical Engineering

The first 150 years of the Institution of Mechanical Engineers

John Pullin

Foreword by HRH The Prince of Wales

Quiller Press
London

All famous engineers and Presidents of the Institution. Left to right: John Penn, Joseph Whitworth, Robert Napier and William Fairbairn

First published 1997 by
Quiller Press Ltd
46 Lillie Road
London SW6 1TN

Copyright © 1997 The Institution of Mechanical Engineers

ISBN 1 899163 28 X

Jacket designed by Roger Fuller
Book designed by Jo Lee
Printed in Hong Kong

Contents

As an Honorary Fellow of the Institution of Mechanical Engineers, I am delighted that this book celebrates not only the 150th anniversary of the Institution, but also chronicles the remarkable progress that has been brought about through the work of mechanical engineers.

Progress is the very spirit of engineering, and the mechanical engineers of the past 150 years have been responsible for innovations that have transformed standards of living and ways of life throughout the world. The contrast between the world today and the world of the founders of the Institution in 1847 is startling.

But today's increasing pace of technological change puts even greater pressure on the problem-solving and innovative capabilities of mechanical engineers. Developments in agriculture and medicine, (and the ethical considerations they raise) the need to conserve the earth's natural resources and to reverse the damage done by over-exploitation, all present tremendous challenges. The demand for engineering innovation and indeed, appropriateness is perhaps greater than ever before.

Pam Liversidge
President 1997/98

Ernest Shannon
President 1996/97

Preface

Pam Liversidge and Ernest Shannon

On January 27, 1847, a group of railway engineers met in Birmingham to establish "a highly respectable Mechanics' Institution". Its purpose was stated simply as "to increase knowledge and give an impulse to inventions likely to be useful to the world" and George Stephenson was elected as its first President.

So progress has been at the heart of the Institution of Mechanical Engineers ever since its birth. In 1920, in the strange, dislocated, period after the First World War, one of our distinguished predecessors as President, Captain H Riall Sankey, returned to such fundamentals. "What can you do without the engineer? – absolutely nothing at all," he said. In more recent and robust times, our trademark has become "Nothing moves without mechanical engineering."

The history of the Institution of Mechanical Engineers over its first 150 years is a story of progress and genuine achievement. The great names of mechanical engineering, Stephenson, Whitworth, Armstrong and Whittle, are all there. But engineering has always been about teamwork, and the exploits of the unsung, unknown teams of mechanical engineers are also acknowledged. Their innovations and their developments are told through the story of their Institution.

The path of progress has not been smooth, and the Institution, like the careers of its members, has had its ups and downs. Today, with more than a thousand times the membership of one hundred and fifty years ago, we can pay tribute to the vision of progress that sustained our founders and their successors. Tomorrow's world of faster transport, easier communications, miniature and intelligent machines, improved life expectancy, extra-terrestrial travel, and much more besides, can only be guessed at. In celebrating 150 years, we look forward with confidence, knowing that the mechanical engineering profession still has a unique contribution to make to the progress of the human race.

June 1997

George Stephenson: the painting of the first president of the IMechE by John Lucas.

1 The Beginnings of a Profession

TO 1847

The Institution of Mechanical Engineers was born in a single evening. The engineers and mechanics who met in the Queen's Hotel, Birmingham, on the evening of Wednesday January 27, 1847, completed their business formalities quickly. They were then able to sit back and listen to the reminiscences of George Stephenson, their newly-elected president and the man whose personality and career had been the inspiration for the new Institution and for the whole profession of mechanical engineering.

The events that led to that Wednesday evening in Birmingham have been the subject of myth, and at times mischief, over the 150 years since. George Stephenson's role in the formation of the IMechE in particular has been the subject of dispute.

Break of gauge: how the Illustrated London News *saw the problems of travelling to Birmingham in 1846 via Gloucester*

One fact that is certain is that, in the early autumn of 1846, an informal discussion between six, maybe seven, men – not all of them engineers – ended with a decision to canvass support for an Institution for "mechanics and engineers".

Exactly where this discussion took place is one of the areas where

James Edward McConnell: the instigator and leading spirit of the IMechE in its early days

myth appears to have crept into the IMechE's history. At the opening of the present headquarters in Birdcage Walk, Westminster, in 1899, the commemorative pamphlet presented to members referred to a meeting of some kind in the house in Cecil Street, Manchester, of Charles Beyer, the manager of Sharp Brothers' locomotive works. Beyer was certainly a prime mover in the formation of the IMechE, but it is likely that this "meeting" was no more than a conversation, though perhaps the idea for the new body was first floated in Manchester.

More probable as the venue for the discussion that led to the first meeting was the Lickey incline near Bromsgrove on the Bristol and Birmingham Railway. James McConnell was, until 1846, locomotive superintendent of this line, known earlier as the Birmingham and Gloucester Railway. McConnell appears to have invited several engineers, including Charles Beyer, and others to observe locomotive trials at Lickey, where the 1 in 37 gradient remains one of the steepest on the British rail network. In one account, which appears to have been written down not at the time but many years later, a shower of rain sent the party scurrying for shelter, which they found in a trackside platelayers' hut. And it was in this hut that the discussion may have turned to the formation of an Institution for mechanical engineers.

The rain shower and the hut are very probably mythical – a pity, because there is undeniable Abraham Lincoln-style appeal in a progression that starts in a line-side hut and finishes at one of the grandest addresses in London.

Instead, it is more likely that the party of engineers repaired to James McConnell's house at Blackwell, less than a mile away, and the idea for a new Institution took shape as the hospitality flowed. Again, though, the process by which the idea was formed is the subject of later mythology. In his biography of George Stephenson, written more than a decade later, Samuel Smiles, the writer who created the Victorian ideal of the self-made man, suggested that the IMechE was formed out of a sense of justifiable rage.

McConnell's house at Blackwell, near the Lickey incline, where the idea for the Institution was probably born. The house has since been demolished

Never one to let the facts get in the way of a good story, Smiles wrote that the engineers who gathered at the Lickey incline were angry that Stephenson, the most famous mechanical engineer of the age, had been refused admission to membership of the Institution of Civil Engineers unless he sent in "a probationary essay as proof of his capacity as an engineer". Stephenson, according to Smiles, had declined to submit to this indignity, and right-thinking engineering opinion was behind him – to the extent of forming an engineering Institution that would not merely accept Stephenson, but would place him at its head.

It took almost a century to expose Smiles's account of the formation of the IMechE as myth – or, at the very most, exaggeration. Researches at the Institution of Civil Engineers in the 1950s, after the centenary history of the IMechE in 1947 had perpetuated the story, discovered the truth. There had been, in the 1820s, definite coolness between the abrasive Stephenson and some prominent members of the Institution of Civil Engineers, and Stephenson seems to have retained a degree of distaste for London-based consulting engineers, as distinct from "practical" northerners.

But the ICE could find no evidence that Stephenson had ever applied for membership, and none that his application had been refused. Nor was there any evidence that the words attached to the refusal, that Stephenson was "merely a blacksmith", had ever been used. Even if some of the leading ICE members reciprocated Stephenson's cordial dislike, the supposed slight has no factual basis:

it appears to have been invented by Smiles some years after Stephenson's death as an illustration of the obstacles that the engineering establishment put in the way of the self-made hero. It also seems to have had greatest currency at the occasional times when relations between the two institutions, usually close, were uneasy.

Samuel Smiles's purpose in embellishing his biographies of Victorian engineers in this way – there are similar distortions in other studies – was his wider political agenda extolling the very Victorian virtues of self-improvement and individual effort. Engineers owe Smiles, himself a one-time railwayman, a great deal for bringing the profession to popular attention and for using engineering careers as exemplars of the ideal of self-help. Modern engineers may feel, though, that his misrepresentations left room for later mischief. They may also feel that one of the legacies of his emphasis on the work of the heroic individual has been a lack of appreciation of the teamwork essential to most branches of engineering and an invidious comparison between the engineers of the past and those of the present day.

The formation of the IMechE in 1846/47, in fact, is an early example of unsung engineers working as a team towards a common purpose. None of the engineers who met at the Lickey incline achieved sufficient individual fame to merit a Smiles biography. None of them, in fact, achieved the accolade of the presidency of the Institution they created.

Apart from McConnell and Beyer, three others are known to have been present. They were Richard Peacock, at this time locomotive superintendent of the Manchester & Sheffield Railway, later to be Beyer's business partner and, later still, a Member of Parliament; and two engineers from the Smethwick works of the Birmingham Patent Tube Company, George Selby and Archibald Slate. Slate was nominated to be the secretary to the group. Also in the party was Charles Geach, a Birmingham banker.

The result of the gathering at Lickey was a round-robin letter to prominent engineers across Britain. It was headed "Institution of Mechanical Engineers".

The original round-robin letter sent in 1846 to engineers inviting them to a meeting to set up the IMechE

It read: "To enable Mechanics and Engineers engaged in the different Manufactories, Railways and other Establishments in the Kingdom, to meet and correspond, and by a mutual interchange of ideas respecting improvements in the various branches of Mechanical Science to increase their knowledge, and give an impulse to inventions likely to be useful to the world. We hope to have the pleasure of seeing you at a Meeting of the promoters of the above on Wednesday 7th October at 2 p.m. at the Queen's Hotel, Birmingham."

The letter was signed by McConnell, Beyer and Slate, and also by Edward Humphrys, of the firm of Rennie's in London. Humphrys had not been present at the gathering at the Lickey incline. The inclusion of his name suggests that the idea for the new Institution had been discussed before; it also gave the letter the endorsement of a London engineer to add to that of the Birmingham and Manchester men, and Rennie's was a lustrous name to attach to the new body. The company had been founded by the engineer John Rennie, who had early mechanical engineering experience with the Boulton & Watt company before becoming one of the great civil engineers of the era of canal and bridge building. Rennie's most famous achievement was the rebuilding of London Bridge, though by 1846 his sons had redirected his company towards marine engine manufacturing.

The round-robin letter appears also to have been carefully phrased. By including mechanics as well as engineers, the new body was giving itself a wider remit than the established Institution of Civil Engineers; putting manufacturing before railways was politically astute, since the pace of railway development and the perceived pushiness of railway engineers was not universally liked; and the references to mechanical science and to inventions made sure that the new Institution was differentiated from the large number of mechanics' institutes that had been set up in the previous 25 years.

The Queen's Hotel in Birmingham, where the preliminary meeting of the IMechE was held on October 7, 1846, formed the entrance to the Curzon Street station which was the first terminus of the London & Birmingham Railway. The hotel was later converted into the London & North Western Railway's goods offices when the passenger traffic moved to New Street.

The Queen's Hotel at the Curzon Street terminus of the London & Birmingham Railway: venue for the inaugural IMechE meetings

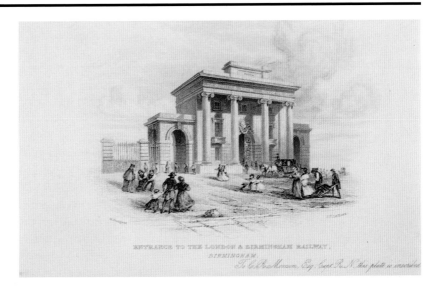

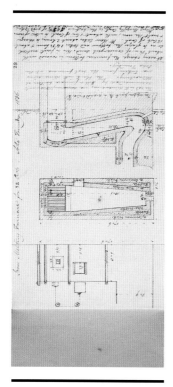

Notebook of the Soho Works, Birmingham, kept by Henry Wright in the 1830s and 1840s. Wright was an early member of the IMechE – though not a founding member – and went on to be a carriage builder in his own right

McConnell chaired the meeting: he was in the throes of moving to become the locomotive superintendent for the southern section of the London & North Western – the old London & Birmingham – based at the Wolverton works, and was therefore the meeting's host as well as its chairman.

Many of the engineers who attended appear to have come from the Birmingham area, and the Boulton & Watt company – another "good" name for the new Institution to have on board – had several representatives. When McConnell was officially elected chairman by the meeting, one of the Boulton & Watt men, William Buckle, the manager of Boulton's Birmingham Mint and later of the Royal Mint, was elected vice-chairman. Again, this was an astute move. Most of the founding fathers of the Institution were relatively young men in their thirties. Buckle, 20 years older, added a useful gravitas.

The October 7 meeting was not composed entirely of Midlands engineers, though. Humphrys, who had signed the letter calling the meeting, came from London; a more significant Londoner, perhaps, was a member of the Cubitt family, Benjamin Cubitt. The Cubitts were one of the greatest of the Victorian contracting dynasties and were prominent in the Institution of Civil Engineers. Benjamin was locomotive superintendent of the South Eastern Railway from 1842 to 1845 and was involved with the civil engineering members of the family in the construction of the Great Northern Railway – the present East Coast mainline – at the time the IMechE was founded.

There were engineers from the north, too. Beyer and Peacock came from Manchester, as did John Ramsbottom, the locomotive engineer at the London & North Western Railway's satellite works at Longsight. Henry Dübs, from Tayleur's Vulcan Foundry at Warrington was, like Charles Beyer, a German emigré who had arrived in Britain to study British engineering and production techniques and had stayed.

This preliminary meeting appears not to have taken too long. The four signatories to the letter of invitation, plus Peacock, Buckle and two more Midlands engineers, John Edward Clift of the Staffordshire and Birmingham Gas Works and Edward A Cowper of Fox & Henderson's Smethwick factory, were elected to form the committee to draft the rules and regulations. Archibald Slate was confirmed as the honorary secretary.

So far, the business of the new Institution had been conducted with speed and high-minded seriousness. That there was also to be another, more sociable, side to the Institution of Mechanical Engineers is indicated by the report of the dinner that followed this meeting in October 1846. The list of toasts suggests that the Samuel Smiles image of the earnest Victorian engineer is not the whole picture.

The revels started with a loyal toast to Queen Victoria and the Prince Consort, but that was a preliminary to a long list of further toasts. The first one was to the success and prosperity of the Institution, coupled with the name of Slate, the secretary; this was followed by a toast to the Institution of Civil Engineers, to which Cubitt replied. From these beginnings, there is a strong suspicion that the evening slid into genial, less than sober, sentimentality. There were toasts to the memory of James Watt; to the Rennie company; to the "locomotive manufactories" of Britain; to the freedom of the press; to the health of George Stephenson and of Robert Stephenson, "the worthy son of a worthy sire"; to the railway directors of Britain; to the health of Isambard Kingdom Brunel; and finally to the chairman James McConnell and the vice-chairman William Buckle. Each toast demanded a response. There is no record of the time the dinner finished, nor of the eventual condition of the participants. The indications are, though, that present-day engineers who have been known to complain about the length of some of the formal proceedings at Institution functions are getting off extremely lightly in comparison to their Victorian forebears.

A little over three months later, on January 27, 1847, the engineers were back at the Queen's Hotel for the formal inauguration of the Institution of Mechanical Engineers. In the intervening period, the committee chosen at the preliminary meeting had met in November to approve the rules and regulations.

Exactly who drafted the rules and regulations, and who co-ordinated the large amount of correspondence that the setting up of the new Institution entailed is not clear. It seems certain, though, that the bulk of the work was done by the five Birmingham members of the committee – McConnell, Slate, Buckle, Clift and Cowper. The committee rules adopted at the October 7, 1846 meeting allowed for a quorum of five out of the eight committee members. The committee also co-opted Charles Geach, the banker who had accompanied the original party to the Lickey incline, as treasurer, and they appear to have made full use of the postal service, which had been made reliable and quick with the growth of the railway network.

The rules were uncomplicated. Membership was open to mechanics and engineers over the age of 21 who were "managing heads of establishments where engines or machinery were made or employed", but this qualification was interpreted rather more liberally for those who styled themselves as engineers than for mechanics, and the two terms were already far from synonymous. The initial list of members was by invitation, but after the January 27, 1847 meeting new members had to be sponsored by three existing members. All members had to pay a first annual subscription of £5, which was the normal £3 subscription plus a £2 joining fee, and a condition of membership was that they should present at least one paper a year to the Institution. This condition was never enforced and it is difficult to see how it could have been, since there were intended to be only four meetings a year at which papers could be read. It was quietly dropped in 1855. Honorary membership was available for eminent people who were not otherwise qualified, and the first eight honorary members were introduced in 1848.

The committee did its work well. By the time of the formal meeting to inaugurate the Institution, 56 had been signed up for membership. They came from all over Britain and Ireland, with one, a Matthew Leahy, from Boulogne in France; many were also members of the Institution of Civil Engineers. The inclusion of one name, "William Lee, Sharebroker, Cheltenham", indicates that membership rules were flexible to a degree.

The committee's masterstroke, though, was to invite George Stephenson to be the first president. Stephenson was 65 and was living in comfortable semi-retirement at Tapton Hall, near Chesterfield, doing some consultancy work, some of it overseas, and dabbling in scientific experiments. Not all of his scientific work was to do with engineering. His Derbyshire neighbour was Joseph Paxton, the man who had been head gardener to the Duke of Devonshire at Chatsworth and who would be the designer of the Crystal Palace for the Great Exhibition of 1851, and Stephenson appears to have tried to rival Paxton in producing fruit and vegetables. A particular concern was trying to persuade cucumbers to grow straight, a problem he solved by enclosing them in glass cylinders. He also experimented with manure, the artificial incubation of eggs and the quick fattening of chickens, anticipating food technologies by several decades.

There is some evidence that Stephenson was initially far from clear what he was taking on with the presidency of the IMechE. He wrote to a friend that he had been invited to be president of "a highly respectable mechanics' institute in Birmingham". The same letter also states that he had turned down pleas to become a Fellow of the Royal Society and a member of the Institution of Civil Engineers, and as there is no record of either body asking Stephenson officially to join, it is possible that he was a little confused.

To Stephenson's credit, though, once the IMechE had been formed, he was wholehearted in his support and regular in his attendance. The role of patriarch suited him – it had been remarked before that, though often blunt in speech, awkward in written communications and brusque in his personal relations, Stephenson appeared to have an affinity with younger engineers, willing to show inordinate patience in dealing with their questions.

The meeting on January 27, 1847 at which the IMechE was formally launched was stage-managed for Stephenson. Again, James McConnell took the chair. He read out a long list of names of the engineers who had responded to the invitation to become members, and then moved the proposal "that the Institution be established". Buckle seconded the proposal, and the resolution was passed. Slate then read out the rules, and George Stephenson, seconded by Beyer, moved their adoption. McConnell then proposed that Stephenson should be appointed president, and the proposal was carried "by acclamation".

The Institution of Mechanical Engineers was on its way, and the most famous mechanical engineer of them all was at its head.

* * * * *

If the birth of the Institution of Mechanical Engineers was a relatively swift affair, the gestation of the profession that it serves was very long indeed.

Mechanical engineering has existed since prehistoric times. Basic mechanical devices – the wheel, the lever, the pulley – predate written history. The earliest civilisations harnessed the natural power sources of water and wind to their own needs, and made tools from minerals and metals. The degree to which early civilisations used engineering to make artefacts and to increase their own capabilities is the primary measure of their civilisation: without engineering, there is no civilisation.

Fundamental mechanisms also go back a long time. Clocks of many kinds and varying accuracies have been known for many centuries, and complex tools for manufacturing precision components in scientific instruments, watches, and locks were in widespread use from the 16th century, in many cases anticipating larger scale engineering production technology by many years. Some of the principles of mechanics, physics and other engineering science were also known from way back, though knowledge did not necessarily mean understanding.

Mechanical engineering from the era before the term was used: pictures of 16th century lifting devices from a book published in 1578

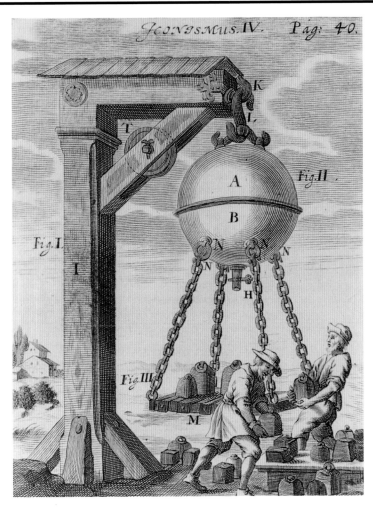

Despite these roots, mechanical engineering as a distinct discipline, as a career and as a profession goes back at the most 200 years, and engineering as a whole is identifiably the separate discipline that we know today for perhaps only a century before that. The ancient Britons, the Romans, the Anglo-Saxons and the Normans had all made indelible imprints on the British landscape through works of civil engineering and construction. But the development of engineering in Britain across the 18th century was on a wholly different scale and pace.

Engineering as a separate discipline fuelled the industrial revolution in Britain from the mid 18th century onwards – and fed off it as well. Engineering emerged as part of, and then contributed to, the series of interdependent events and progressions which made up the transition from an agrarian economy to an industrial one. Economic and social historians disagree about the timing and the causes of the industrial revolution, and many of them argue that the word "revolution" is a misnomer.

However, virtually all agree that, from around the middle of the 18th century, there was a change in the British economy and in British society: before 1750, though there had been undeniable progress in terms of artefacts and living standards, Britain was still basically an agricultural country with a land ownership and political system not much advanced from medieval times. After 1750, Britain starts to become an industrial country and the rate of change accelerates dramatically, with effects on the whole population.

More than that, though, the changes of the industrial revolution were self-sustaining and self-fuelling. Innovations in one area of activity led to innovations in other areas. Economic upheaval such as the growth of companies and of trade created further upheaval which made companies and trade grow faster. Agricultural improvements displaced many people from the land into the new urban communities: the towns then needed to be fed, so further agricultural improvements were needed.

Engineering was fundamental to the industrial revolution. In philosophical terms, engineering embodied 18th century views about the power of reason and the confidence that man could, to a degree hitherto not attempted, impose order on the chaos of nature and improve upon the natural world. Great civil engineering works such as John Smeaton's Eddystone lighthouse or the canals that were built across Britain from the 1750s onwards were expressions of this confidence; so, too, in a different way, were the great formal gardens of the period before Capability Brown.

In parallel with this confidence to subjugate nature was the notion of extending mankind's own natural abilities in terms of strength, speed, stamina, precision and awareness of the wider world. A broad difference between the machines developed before the industrial revolution and those developed during and after it is that the earlier tools and mechanisms assisted existing capabilities, whereas the newer devices allowed tasks to be performed that could not have been done before and which were outside the range of man's own abilities. The scale of ambition is different.

Central to both the philosophy and the ambition is the idea of progress through innovation, and the industrial revolution was built on fundamental engineering innovations in technology and in materials.

Falteringly, from the beginning of the 18th century and the work of Savery and Newcomen, later followed up and developed by James Watt, ideas were gathering about the development of power sources over and above those provided in nature. Early steam engines were stationary devices – it took Richard Trevithick and then George and Robert Stephenson to attach the steam engine to the idea of self-propulsion, or locomotion – but even stationary steam engines were portable in a way that water power was not. This portability would, eventually, have a big effect on industrial geography.

The primary early use for reciprocating steam engines of the kind built by Newcomen was in pumping, and particularly in extracting

Jonathan Hull's steam tug patent of 1736 proposed using a Newcomen engine to pull ships in and out of harbour

water from coal mines and other underground mineral workings. Coal was the fuel that powered the industrial revolution and the development early in the 18th century by the Darbys of Coalbrookdale in Shropshire of coke-smelted iron provided the basic raw material of the industrial revolution. The interdependence of the steam engine, coal and iron dates from the beginnings of industrialisation: Abraham Darby's coke-smelted iron of 1709 was used to build Newcomen engines that, three years later, would find their first true application pumping out mine workings that would provide Darby with the fuel for his process.

The industrial revolution is full of interdependencies like this: the complexity of trying to sort out the real sequence of events, few of which were written down at the time, is one reason why no two histories of the period tell the same story.

Innovation was, in any case, a fairly haphazard process throughout the 18th century. Information about technical developments had no regular or reliable outlet. There were no technical journals, and, until in 1771 Smeaton and his friends formed an unofficial discussion group which later, after his death, took on a more permanent form, there were no specifically technical or industrial societies either. Bodies such as the Royal Society, and the later Royal Society of Arts, concerned themselves to a degree with what we would now recognise as engineering, and the Philosophical Transactions of the Royal Society contained important early papers on steam. Smeaton himself contributed papers to the Royal Society, which was also important in bringing in Continental European experience. But these societies were very much part of a scientific establishment and an amateur tradition: the connection between science and engineering was far from clear, and the connection between science and manufacturing even more tenuous.

A further hindrance to the dissemination of information was that travel was, until the last third of the 18th century, a painfully slow affair, so developments in one region did not necessarily reach wider attention with any speed.

The result was that there was often a long timelag between invention and widespread adoption of both technologies and products. Darby's invention of coke-smelted iron might have been expected to

The textile industry was the first to embrace mechanisation: this is Arkwright's carding machine of 1775

Richard Arkwright's business card from the time before he set up his textile factory at Cromford in Derbyshire

change the whole economic base of the iron industry in Britain and beyond since, suddenly, the need to fell vast tracts of forest to provide the charcoal for smelting had disappeared. In fact, coke smelting was scarcely used outside Coalbrookdale until the 1760s, taking 50 years to reach even the Black Country, 20 miles away. The last ironworks to use charcoal for smelting survived well into the 19th century.

The problems of disseminating information about inventions in the 18th century were matched by an almost total lack of information at the time about incremental innovations – the gradual improvements to technologies, devices and methods – and the shortage of information in this area persists to this day. It is known, for instance, that the efficiency of the basic Newcomen reciprocating steam engine was increased by three or four times through incremental innovations over the first two-thirds of the 18th century, but where, when and how are far from clear. The fact that Watt's steam condenser and rotary engine then doubled engine efficiency again takes on less significance in the context of this earlier improvement.

In fact, industrialisation and engineering innovation in the industrial revolution are as much about this kind of continuous improvement as they are about the great inventions, and there is no reason to suppose that teamwork, rather than individual inspiration, was any less important to early mechanical engineers than it is to today's engineers. Particularly, the adoption of machinery in different sectors of industry, which was a large part of the genesis of the separate profession of mechanical engineering, was a gradual process with few identifiable benchmarks.

The textile industry is recognised as being the first to mechanise and to adopt the factory system of working, though in practice it was the spinning side of textiles only which became fully mechanised in the 18th century, with steam power then applied from the 1780s onwards. Pioneers like Richard Arkwright at Cromford in Derbyshire sited their mills next to fast-flowing water, and dug canals both to ensure water supplies for the power and the process and for transport.

As the use of steam power progressed, the need for water as a power source declined and the industrial geography of Britain changed. The cotton trade had been concentrated in Lancashire since the 17th century because of its nearness to the port at Liverpool where the raw material was imported; with the coming of steam power, the spinning side of the industry moved down into the lowlands where there was still sufficient water for the process, but water power for the mills was not required. Other industries located themselves close to sources of iron and coal, or close to where the customers were. The legacy of the geographical shifts brought about in the industrial revolution is still apparent today, with concentrations of engineering industry in the West Midlands, Lancashire, Yorkshire and the West of Scotland.

Steam power had a further effect on mechanisation. Early machinery in the new factories of the textile industry and elsewhere used wood and stone as base materials, as milling machinery had for centuries. The stresses and strains imposed by steam power on machinery was altogether too much for the traditional materials to handle: in another example of the interdependencies of the industrial revolution, steam engines, themselves made of iron, forced the introduction of iron across the range of industrial machinery.

The widespread adoption of iron as the basic material of industrialisation and mechanisation was itself a key stage in the development of mechanical engineering as a distinct discipline in its own right. Some of the credit for this stage is probably due to James Watt and his partner, the Birmingham ironfounder Matthew Boulton.

It was the Boulton & Watt company that appears to have been among the first to bring together the traditional machine-building skills of the millwrights with the precision engineering skills of the scientific instrument makers and the clock and watch makers, who had previously worked mostly in brass.

The bringing together of these two sets of skills created precision machinery on a large scale in a new material. Watt's steam engines were the power sources for the mechanisation of other industries: in this instance, though, they were also the instruments by which the profession of mechanical engineering developed.

The recruitment by Boulton & Watt of large numbers of artisans, both millwrights and instrument makers, is important on another account. It indicates also that mechanical engineering was, from its beginnings, an unusually open profession. What were essentially the craft skills of metalworking were harnessed to new technologies in materials and machines to create the new discipline.

The precision skills of the locksmith were applied to the new material, iron: Bramah's lock was the state of the art for 50 years to 1851

Of all activities and professions, engineering appears to have been from the outset the least affected by class and education. The church and the law were dominated by the aristocracy and the landed gentry; art and literature similarly remained the preserves of the well-to-do; political influence, including the basic right to vote, was inextricably bound up with the ownership of land, and that was in the hands of a small fraction of the upper classes. Engineering by contrast put the skilled artisan alongside the scientifically inclined aristocrats in positions of some equality: in some respects, more than equality, since the engineering profession in Britain developed very much as a practical response to problems, rather than as a scientifically based discipline.

The great engineers of the industrial revolution come from a variety of backgrounds unthinkable in any other sphere at the time; some of them, George Stephenson included, had very little formal education. There is a strong connection between non-conformists, who were often denied education in England, and the engineers of the Industrial Revolution. Engineering was the first of the professions to offer a degree of social mobility which was not entirely related to wealth, though many successful engineers did become rich.

Engineering in Britain had other distinctive features. One reason why it was, from the earliest days, the most meritocratic of professions was that it developed as a practical subject in which knowledge increased and progress was made through experimentation and other empirical methods, rather than by the application of scientific knowledge and theory. Much of the underpinning science which is today regarded as essential learning for engineers of all types was not known at the time of the industrial revolution. Even where it was known it was often little understood. The steam engine pioneers, incredibly to modern thinking, had no conception what heat was; they knew that their engines worked, but their explanations of why they worked are often wrong.

The British engineering tradition of development through experimentation is one reason why the British engineers of the 18th and 19th centuries appear extremely creative in terms of innovation. Some historians argue, however, that the lack of a scientific foundation in British engineering may have been a factor in the erosion of the industrial lead in the second half of the 19th century, the haphazard record of innovation in Britain contrasting with a more planned industrialisation and application of new technologies in other countries.

The lack of scientific grounding for engineers extended to the education system, though the stricture applies much less to Scotland than to other parts of Britain. Scottish universities started teaching mechanical science as part of natural philosophy in the mid 18th century, and James Watt and other great Scottish engineers had some science training. The liberal Scottish education tradition is an important factor in industrialisation across the whole of Britain.

But south of the border in England and Wales, engineering as an academic subject made no sustained impact until well into the 19th century. The first professor of engineering at an English university was appointed only in the late 1820s; the first professor of mechanical engineering took up his post in the same year, 1847, that the Institution of Mechanical Engineers was formed. Even when engineering was admitted as a subject fit for study in the mid 19th century, some academic bodies found difficulty categorising it and placed it in the arts faculty.

Engineering in the early part of the industrial revolution was primarily about solving problems, and the engineering tradition that was created was one of immense practicality, of making things work.

Many of the famous names of 18th century engineering were generalist problem-solvers, rather than specialist innovators. They expropriated the word "engineer" from the military, where it had been used since medieval times to denote the men who built engines of war. Sir William Fairbairn, third president of the IMechE after the Stephensons, claimed in his autobiography that John Smeaton had been the first to style himself as an engineer while working on a canal in Staffordshire in 1761.

Men such as Smeaton and the canal-builder James Brindley owed their fame to the broad vision and the unprecedented scale of their works, as much as to any new technologies used. The great legacy of these early engineers was the canal system. Canals, though, were not new: rivers had been canalised for centuries before, and artificial channels had been dug, in the Fens, for instance, for drainage. The difference that the 18th century engineers brought was the scale of their vision of a network of canals covering all of mainland Britain, the linking of that vision to the needs of industry for transport, and the entrepreneurial and organisational ability to carry it all off without losing a fortune.

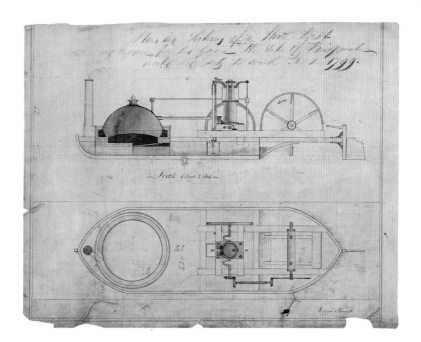

William Sherratt's drawing of a steam boat for the Duke of Bridgewater's canal dates from 1799

The tradition of generalist engineers was continued by Thomas Telford, originally a stonemason, and, into the 19th century, by Isambard Kingdom Brunel. Telford, never short of self-confidence, certainly considered himself qualified to tackle problems across the range of engineering, from beam engines to bridges, using materials from stone to iron. As one of the prime applications of early steam engines was in pumping water for the canals, he had plenty of opportunity to demonstrate the breadth of his engineering talents.

By the end of the 18th century, however, the rise of the mechanised factory as the basic unit of industrial organisation had increased the range of skills that were demanded within engineering so that the

Isambard Kingdom Brunel at the shipyard building the Great Eastern in 1857. Brunel is second from the right in the foreground; the ship's builder, John Scott Russell, is standing on the left, just back from the foreground figures

The portrait of James Watt by Frederick von Breda hangs at IMechE headquarters

single generalist discipline was starting to fragment into several component parts. Mechanical engineering, as a separate discipline with its own issues and priorities, probably dates from James Watt, who was one of the first of the great engineers to devote himself almost exclusively to the development of machinery, and particularly the steam engine.

Watt is not an easy character for today's engineers to warm to. His development of the reciprocating steam engine made possible the mechanisation of large parts of manufacturing industry. He was responsible for defining one of the basic units of engineering, the horse-power, which usefully related the unfamiliar machine world back to the language that all could understand. His alliance with Matthew Boulton brought about the marriage of the large-scale metalworking and mechanical skills with those of the precision instrument makers, and formed the foundation of mechanical engineering. His work, drawing from his Scottish background in Clydeside shipbuilding and his contact with Scottish engineering pioneers such as Roebuck and Murdoch, was helped by developments in materials by men such as Darby and Henry Cort. From it, the iron and steel age begins.

Yet Watt, with Boulton's encouragement, also tied up much of his work in patents. This helped to stifle innovation for many years, and then had the contrary effect of making many later engineers wary of patenting their inventions, though in many cases, such as Brunel, the wariness was less to do with high-minded sentiments about developments being for the public good and more to do with the probability that the publication of a patent would give too much away to rivals.

Watt engine of 1795 for Messrs Grimshaw

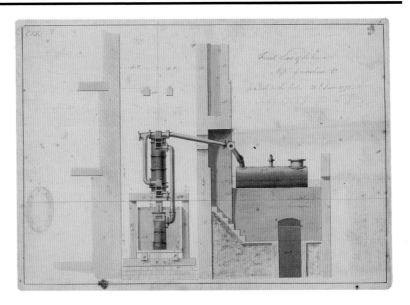

James Watt's snuffbox, now held in the IMechE archives, resting on a letter signed by Watt

Watt was also personally discouraging to other engineers who wanted to take the steam engine on into new applications, including transport. Intensely conservative and constantly grumbling about his headaches, Watt cuts a less than sympathetic figure. He appears to have even needed to be prodded and cajoled by Boulton into developing the separate steam condenser that increased the efficiency of Newcomen's engine and made rotary motion possible.

It is easy to overestimate the extent of 18th century industrialisation and mechanisation. By 1800, when Watt's patents on rotary motion engines were finally overturned, there were probably no more than 1,000 steam engines in all applications in Britain, and around half of them had been built by Watt. In terms of industrial use, mining and the textile industry, along with brewing, were substantial users, but mechanisation was still in its infancy elsewhere. The acceleration comes after 1800.

The late 18th century, though, had sown the seeds of the Industrial Revolution, and the beginning of the 19th century reaped the harvest. There are three strands to the growth of engineering and the development of mechanical engineering as a separate branch in the early part of the 19th century.

One was the extension of engineering and mechanisation into new areas of industry: the development of a whole new range of technologies and techniques based on the marriage of heavy metalworking and precision engineering and the use of the "new" material, iron. The first 30 years of the century saw the factory system applied to a huge range of industries. It saw the start of the emergence of modern industrial issues, such as the availability of raw materials and labour,

James Nasmyth's sketch of his own works at Patricroft in about 1840

productivity and efficiency, manufacturing quality and standardisa-tion. Engineering and the introduction of machinery were central to the growth of workshops and of companies as the basic economic units of production.

A second strand is the application of engineering outside industry. In the first half of the 19th century, this means principally one thing: the application of Watt's steam engine to transport. Above all, apply-ing mechanical engineering to the railways would be the spur to the most far-reaching social, economic and industrial changes of them all in the years after 1830 – the years immediately before the forma-tion of the Institution of Mechanical Engineers in 1847.

The third strand is the increasing self-awareness among engineers that theirs was a profession with a discrete identity and a distinct body of knowledge, and, increasingly, a profession that had the unde-niable power that comes with economic clout. The first half of the 19th century saw the first engineering institutions, but more gener-ally it was an age of immense growth in engineering education and information, not just for "professional" engineers but also for arti-sans.

The dramatic extension of engineering into new areas of industry was the result of a series of inventions and incremental innovations, many of which are difficult to pin down to a specific date or a specific inventor. It is clear, though, that following on from Boulton & Watt's work in Birmingham, other engineers in London, Manchester and elsewhere were setting up engineering works on the same lines, employing craft skills to make new products. Importantly, Joseph Bramah started an engineering "dynasty" that led on, through Maudslay, to Nasmyth and Whitworth and other engineers prominent in the early days of the IMechE.

Bramah was an example of the way in which engineering offered, from the beginning, a route to respectability for the very poor. His life story reads like an archetypal Victorian morality tale of the kind that inspired Samuel Smiles. Born as Joseph Brammer in Yorkshire and apprenticed as a joiner and cabinet-maker after an injury to a leg

*Joseph Bramah, born Brammer,
was an engineering inventor of
breadth and diversity*

*Bramah's Water Closet Upon a
New Construction was the
source of his wealth*

meant that he was not fit to work on the land, he nevertheless managed to walk to London in search of work. His woodworking business in the capital took him into some of the biggest houses, and he noted that the aristocracy's sanitary arrangements were no less primitive than those of the Yorkshire peasantry. Brammer's name change to the more aristocratic-sounding Bramah, with its long upperclass vowel sounds, came when he patented his first water closet. It made his fortune. By 1797, more than 6,000 had been sold, and Bramah was in business as a manufacturer, with profits to spare to indulge his talents for invention. His assistant was the engineering metalsmith, Henry Maudslay.

Bramah's main contribution to engineering innovation is, in fact, not the water closet but the hydraulic press, which had been described as a theoretical device before, but had not been built successfully. In fact, the success was down to Maudslay, who overcame the problem of sealing the ram, and arguably Bramah's importance is less his own invention – though it led, in time, to a whole new branch of mechanical engineering – and more the influence he, and then Maudslay, had over the rising generation of engineers. Some of that influence came from the work that they undertook jointly into developing locks using iron, rather than the traditional brass; the Bramah lock remained the standard British lock for 50 years.

Maudslay's later influence was, if anything, greater than Bramah's. He founded his own engineering works in Lambeth in 1810 and from then onwards oversaw an explosive growth in manufacturing technologies. Basic machine tools available before 1810 included the boring machine, built first by the Chester iron founder John Wilkinson in about 1775 and later improved by Boulton & Watt. At Lambeth, other metalworking operations, such as lathes and planing machines, were mechanised, with Joseph Clement, Maudslay's partner, later credited with many of the inventions.

Henry Maudslay's screw-cutting lathe of 1800: Maudslay's ideas influenced a whole generation of mechanical engineers

In practice, the development of the techniques of engineering production at Lambeth was more to do with incremental innovation than with invention: many of the machines first used at Maudslay's were adaptations of tools used earlier in smaller-scale scientific instrument manufacture and watch and clock making. The slide-rest, for example, was by no means a new device when, in the 1780s, it was first attached to a lathe by French engineer Jacques de Vaucanson with a manually-controlled screw; Maudslay (or, in some accounts, Clement) may not even have been first to attach the sliding toolholder to the spindle by gearing, since there is some evidence that an American inventor got there at much the same time.

What is not disputed, though, is that through Maudslay ideas like this were spread far and wide. Eminent engineers such as Clement, James Nasmyth and Joseph Whitworth worked as young men at the Lambeth factory and then went to set up their own workshops which emulated and built on Maudslay's legacy of innovation and attention to accuracy. Nasmyth was later the inventor of the steam hammer and other manufacturing machinery as well as being a pioneer of factory and workshop organisation. Whitworth, president of the IMechE on two separate occasions, achieved worldwide renown for taking over and expanding on Maudslay's ideas for accuracy and standardisation of measurements.

Both of these acolytes of Maudslay, though, made their personal reputations and their fortunes not in London, but in Manchester. Manchester's pre-eminence in engineering in the first half of the 19th century was the result of the early mechanisation of the textile industry. The textile mills of Lancashire and Yorkshire were the first substantial users of machines, and the pace of mechanisation accelerated after the introduction of power looms, invented in the 1780s by Edward Cartwright, but not in widespread use until the 1820s.

The Manchester machinery-building business created by textiles was both innovative and expansive, introducing new technologies in manufacture and interested in new markets. It is no coincidence that the first important steam-powered railway, the Liverpool & Manchester Railway line built by George Stephenson, should have been constructed to give Manchester's manufacturers, particularly textile mills but also others in different markets, quicker access to the port of Liverpool.

The construction of the Liverpool & Manchester is an indication of how much the economic fundamentals had changed in Britain in the short space of a couple of generations with the development of the first national transport systems.

The canals were the first truly national system of transport intended for moving goods – bulk materials and manufactured articles. The period of rapid development of canals started from the late 1750s. The first artificial waterway in Britain is usually given as the Bridgewater canal that brought coals from the Duke of Bridgewater's mines at Worsley into Manchester from about 1767; in fact, there were earlier canals in Staffordshire, and sections of many rivers had been canalised earlier in the 18th century.

Until the growth of the canal system, the businesses of manufacturers and minerals producers were heavily influenced by geography. Unless they were sited close to the sea or near navigable rivers, their trade was likely to be confined to a local area. There were exceptions, of course, where there were long-standing national or international links, but in these cases, trade tended to be continuous and was rarely subject to much competition.

The dramatic growth of transport, starting with the canal system, in the second half of the 18th century changed markets for a whole range of goods. It made possible the movement of bulk materials, such as coal and minerals, from one area to another; it stimulated competition in local, national and international markets.

The canal mania of the late 18th century had other effects. The rash of speculative canal ventures produced spectacular business failures as well as successes, but win or lose, the canals were influential in developing ideas about company formation and share holding and in attracting capital into industrialisation. Land owners saw other uses for their land and their money: the canals were fundamental in detaching the notion of wealth from the ownership of land and in convincing land owners that there were other sources of wealth apart from agriculture and rents.

But the canals were not the only transport revolution of the late 18th century. The formation of turnpike trusts dates back to the 17th century, but the real progress came with the technical developments in road surfacing and in bridges introduced by engineers such as Telford and John Loudon McAdam. These brought dramatic improvements in road transport and led to the development of a nationwide system of stagecoaches. Before the road improvements, a day's journey might be no more than 15 or 20 miles. Parson

Woodforde's diary, written in the 1740s, recounts a three-day expedition from Bath to Oxford: the first day's travel took the party little beyond what is now the motorway junction for Bath on the M4. The stagecoach network meant that this kind of journey could be accomplished in less than a day. As late as the 1820s, though, longer journeys, between Scotland and the south of England, for example, took several days and often entailed a complex combination of sea voyage and land transport.

As with mechanisation in industry, though, the real progress in transport during the industrial revolution came in the first half of the 19th century rather than through the developments of the late 18th century. Again, it relied on an intensive period of mostly incremental innovation.

The idea of railways was not new: horse-drawn wagons had been running on wooden rails in mines and quarries and on harboursides for several centuries. Two sets of innovation transformed railways from just another localised method for moving bulk loads into what was to become, in the space of fewer than 40 years, the great engine of progress in mechanical engineering.

One was the series of developments in the iron industry, starting with Abraham Darby and culminating in the 1770s and 1780s with Henry Cort's technologies for making large-scale iron products, which made the iron rail practicable. The other, more fundamental, innovation was locomotion.

The application of the steam engine to self-propulsion was the result of a series of developments which followed the expiry of Watt's patent in 1800. But before the turn of the century, the Cornish engineer Richard Trevithick had experimented with high pressure engines which increased the power of Watt's basic engine substantially; Trevithick also developed a cylindrical boiler made of cast iron. Watt himself was fairly scathing about Trevithick and his high pressure steam engines, and even suggested that they were criminally dangerous. But the high pressure double-acting engines were successful in mining in Cornwall and in South Wales in the first years of the 19th century.

London demonstration:
Richard Trevithick brought one
of his Cornish locomotives to
Russell Square in 1808

Watt was right, though, to see something of the showman and the adventurer in Trevithick, and the Cornish engineer's later career included several bankruptcies, one of them in Latin America where he invested unwisely in a Panamanian pearl-fishing venture. However, Trevithick's showmanship had much to do with getting the idea of steam locomotion to a wider audience.

Initial experiments in Cornwall in 1801 with a road locomotive known locally as Dick's Firedragon were followed, in 1804, by a London demonstration in which a small loco ran on a circular track at Russell Square. Though the demonstration did not make the inventor the money he hoped for – Trevithick ambitiously charged the general public a shilling for a ride – it did prove the idea of a locomotive with smooth wheels running on a smooth track.

Over the next 20 years, the steam locomotive went from being a curiosity to become a practical proposition for rail haulage. Many engineers contributed to the locomotive's development, but George Stephenson is normally credited with the concept of the railway as a whole. And while it is difficult to identify with certainty any single railway innovation for which Stephenson was individually responsible, there is little doubt that the overall credit is well deserved.

Illiterate until adulthood, Stephenson was self-taught in virtually everything he did, which was one reason why he became such as hero to the Victorian ideal of self-improvement. His own account of his early life was given to the very first meeting of the IMechE on January 27, 1847. He had, he said, "served no apprenticeship to mechanics" and had been rebuffed in every attempt he had made to gain employment in engineering. He had decided to go to America with two friends in the hope of "getting into some Manufactory of Steam Engines", but the friends both got married, and their wives forbade the idea.

Stephenson's opportunity came when a new pumping engine was put into a colliery in the Killingworth group, and proved inadequate to the task. Stephenson had forecast that it would fail, and appears to have been none too reticent about saying that he could remedy the problem. After some weeks, the colliery owner called him in and gave him three days to sort the engine out, which he did with an almost total rebuild. After this, he was put in charge of all the machinery at the group of coalmines centred on Killingworth and he seems to have cemented his position by marrying the mine owner's daughter. As machinery supervisor, he started designing his own stationary steam engines – and was also responsible for the colliery railway systems, which used both horses and the stationary engines for drawing wagons.

The use of locomotives on their railways appears to have been the goal of several of the mine owners in the north-east in the years after Trevithick's London demonstration. High pressure locomotives based on Trevithick designs were built at Gateshead and were used on Tyneside and in Yorkshire, but with limited success. Breakdowns and blow-ups were commonplace.

Then Stephenson was commissioned to build a locomotive for the Killingworth mine. Why he succeeded where others had met with doubtful success is not clear, though he appears to have had an instinctive appreciation of the need for accuracy in both machining

Robert Stephenson & Co's locomotive manufacturing works
at Newcastle upon Tyne was set
up by George Stephenson in
1823 and was one of the largest
producers throughout the 19th
century

Stephenson's friends and rivals:
steam engine innovation moved
in the first 20 years of the 19th
century to the coalfields of the
North East

Stephenson engines at work in
the Northumberland and
Durham coalfields, engraved
by TH Hare. The colliery is at
Hetton

and assembly. The result was the Blucher of 1814, named after the Prussian general, and it was quickly followed by the Wellington and My Lord. By the early 1820s, Stephenson was being commissioned to build locomotives by other north-east mine owners, and in 1823 he set up in locomotive manufacturing on his own account, later taking on his son, Robert, as general manager and naming the company Robert Stephenson & Co.

At this stage, Stephenson's fame in steam engines was confined to the north-east, and he was just one of a number of engineering entrepreneurs across Britain investing in what was still a relatively new technology. He had achieved a wider renown with the invention of the miners' safety lamp, but a parallel development at much the same time by the president of the Royal Society, Sir Humphry Davy, was given precedence by the scientific establishment.

The Stephenson safety lamp. The scarcely-known Tyneside engineer had to share public accolades with the establishment's favourite scientist, Sir Humphry Davy

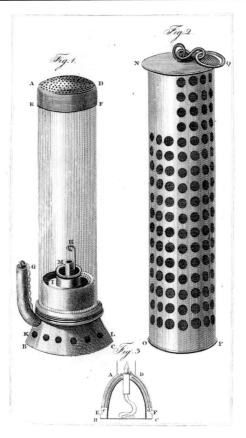

The events that then catapulted George Stephenson to undisputed worldwide eminence, which radically reshaped the whole of the economy and industry and which led indirectly to the formation of the Institution of Mechanical Engineers, took fewer than 10 years.

Stephenson's Stockton & Darlington Railway was built primarily to carry goods: early railways consistently underestimated their appeal to passengers

They started with an extension to his existing work. The Stockton & Darlington Railway on Teesside is often cited as the world's first proper passenger-carrying railway. In fact, it was intended to be nothing of the sort. The railway was promoted by mine owners and businessmen as a method of opening up the County Durham coalfield. The original intention was for a freight-only line, with wagons moved by horses or by stationary steam engines. Stephenson was engaged to supply the engines, and appears to have persuaded the railway promoters to give his self-propelling engines a trial. The result was the engine No 1, later called Locomotion.

From the Stockton & Darlington, Stephenson's reputation spread. But it was his next project, the construction of the Liverpool & Manchester Railway, that changed the world.

As with the Stockton & Darlington, the Liverpool & Manchester Railway was conceived as a freight line. Right up to the time of the formation of the Institution of Mechanical Engineers in 1847, railway builders rarely failed to be amazed by the fact that their railways attracted large numbers of passengers. In the great days of railway building, freight traffic usually provided the economic justification for a new line: almost always as well, freight revenues were overestimated, while passenger traffic was underestimated or discounted entirely.

The Liverpool & Manchester Railway was on a scale not attempted before in railway construction

Stephenson's job on the Liverpool & Manchester was different in many respects from his previous work. The sheer scale of the enterprise was much greater than projects like the Stockton & Darlington. The intention from the outset with a line 31 miles long was that steam locomotives should be the main motive power. As a civil engineering project, with bridges, tunnels and a section of line over a notorious bog called Chat Moss, it ranked alongside the most challenging canal construction projects. To tackle this while introducing the new technology of steam locomotion puts the Liverpool & Manchester into a different league.

Stephenson was engaged as the engineer for the construction of the project; his company was also invited to be one of those tendering for the locomotives, but his main responsibility was to get the line built. In the event, of course, and thanks largely to the engineering discipline which Robert Stephenson brought to locomotive manufacturing in the Newcastle factory, the Stephenson locomotive Rocket was able to win the contest to provide the motive for the line as well.

But George Stephenson's main task, and his big achievement, was the construction of the line. It enabled him to set standards and dictate formats that persist to this day. The standard railway gauge of 4 ft 8½ inches was a legacy from his mine railways; he had been a pioneer, too, of flanged wheels on the wagons, which contrasted with the earlier practice of smooth wheels running within a flanged set of rails.

Perhaps more importantly, though, Stephenson took on, and this time saw off, the engineering establishment.

Part of the planned line ran across the bog of Chat Moss. Stephenson maintained that this section of the line would not be a problem; a survey of the terrain indicated otherwise and the railway directors were persuaded to call in distinguished engineers from London to advise on the construction. The London engineers, with

The treeless waste of Chat Moss, pictured here in the famous Bury drawing, presented one of the big civil engineering obstacles on the Liverpool & Manchester Railway

the survey evidence to back them, were critical of Stephenson, his work and his methods, and there was at least some suggestion of snobbery. Stephenson for his part appears to have treated the London consultants with some disdain. But he managed to convince the railway directors that he knew what he was doing.

Events justified Stephenson's confidence over Chat Moss. More than that, the Liverpool & Manchester Railway was an instant success and opened the floodgates for railway development. Stephenson had succeeded in bringing together the workshop and factory practice of engineering, the harnessing of energy to transport in the steam locomotive and the navvying and construction skills of civil engineering, all in one project. His was a personal triumph of practicality, organisation and ambition. The Rocket captured the public imagination and Stephenson achieved nationwide fame.

Rainhill relic: Stephenson's Rocket, photographed in 1862 when it was sent to the Patent Office for preservation

The railway was also a triumph for the practical and provincial strand of engineering. If Samuel Smiles's biography of George Stephenson made free with the facts about his supposed application to join the Institution of Civil Engineers in London and the rejection of that application, there was certainly some substance to the rivalries between the metropolitan and the provincial engineering traditions, exemplified by Stephenson's spat with London consultants over the Chat Moss section of the Liverpool & Manchester line.

As with the manufacturing side of engineering, which had seen engineers such as Whitworth and Nasmyth leave London for Manchester, many of the real innovations in steam engine development and in transport technology were taking place outside the capital. The impetus for railway development started from the North and the Midlands, with London a little way behind, partly because the original rationale for railways was freight movement, not passenger traffic and the new industries of the provincial cities relied on the raw materials of coal and iron that were mined nearby.

There was a London tradition of engineering, of course. London still had several large engineering employers and was particularly strong until the late 19th century in shipbuilding, but, aside from Maudslay's works at Lambeth and various marine engine works on the Thames, including that of John Rennie's sons at Blackfriars, few of the London engineers appear to have been in the forefront of mechanical engineering technology at this stage.

Instead, the capital, and in particular the area to the west of Westminster, up towards what would later be called Victoria, had developed as the home of the freelance engineering consultants, such as the Brunels and Telford.

Some of these leading consulting engineers, especially Thomas Telford, were, by the 1820s, highly respected and respectable figures. But they were also increasingly conservative. Telford himself appears to have relished the role of sage, and seems not to have been entirely aware that engineering was developing so fast that some of his pronouncements were starting to appear less than wise, and more than old-fashioned. He was, for instance, far from convinced that railways using steam locomotives had any real practical application.

The consultants' work was largely project-based and was international in its scope. These engineers saw no distinction between the construction – and even architectural – activities that today would be broadly classified as civil engineering and the new mechanical engineering and manufacturing businesses that were growing up on the back of advances in engines and machinery. They prided themselves on the ability to provide practical solutions to problems, whether by the design of structures or by the application of machines, and most of them were innovative in the application of existing technologies, rather than being inventive in their own right.

The consultant tradition of engineering in London is a strong and an important one. By the mid 1830s, even determinedly provincial engineers like the Stephensons had joined the brass-plate engineers of Westminster, even if George Stephenson in particular seems to have used the London address only infrequently.

The younger Stephenson, though, and his great friend and rival, the younger Brunel, are in many ways the culmination of this tradition of the generalist consulting engineer. Both Robert Stephenson and Isambard Kingdom Brunel were engineers of vision who extended the range of engineering in terms of both structures and machinery by building bigger and better. Both were innovative across a wide spectrum of engineering, but their innovation was more to do with scale and ambition, with pushing out the boundaries of existing technologies and with novel combinations of known ideas than it was to do with invention. The year that both Brunel and Robert Stephenson died, 1859, is as good a date as any to put on the final transformation of engineering from a generalist to a specialised discipline.

While the London engineers of the early part of the 19th century may not have been leading the way in the technology or the manufacture of machinery, they were important in the development of the engineering profession's identity – and then in the creation of the bodies and organisations that formalised this identity.

There were two stages to this process. The first was the increasing awareness, developing from across the second half of the 18th century, that engineering did constitute a discrete subject, related to, but distinct from, other areas of human endeavour and study. This realisation was, in itself, no mean achievement, given the lack of a scientific basis for engineering knowledge.

The second stage was the inclination, particularly marked in the first half of the 19th century, to form societies and institutions to discuss all kinds of knowledge. These societies were part of wider moves involving the dissemination of information through printed material and were the first tentative steps towards an education system which would take in more subjects than the traditional classics taught to the aristocracy.

The process of forming societies for educational purposes was not confined to the professions and the middle classes: in fact, engineering employers often led the way in providing their artisans and working men with education in arts, literature, science – and engineering matters. Once again, in matters educational, Scotland was ahead. The mechanics' institutes movement had its origins in Scotland in the late 18th century, when the lectures on mechanical science at Glasgow University were opened to craftsmen and other workers "in their working clothes". But it really took off in the 1820s when George Birkbeck moved from the liberal Scottish education tradition in Glasgow to London.

Birkbeck chaired the first meeting of the London Mechanics Institute at the Crown & Anchor tavern on The Strand in November 1823; also at this preliminary meeting were John Martineau, Bryan Donkin and Alexander Galloway, all of them prominent engineering employers in London. These three also provided evidence at about this time to an influential parliamentary select committee on "artizans and machinery" that gave official blessing to the education of working men. Galloway in particular stressed that "a man is not of much use to me unless he can read and write".

The mechanics' institutes movement spread across Britain like wildfire. More than 2,000 artisans and workmen turned up to the second London meeting; within five years, classes, reading rooms and libraries had been set up in most towns and cities. Instruction was given not just on reading and writing, but on industrially useful skills like accurate drawing and measurement, mechanics, mathematics, chemistry, and mechanical and natural philosophy. Directly or indirectly, the institutes provided much of the raw material – in terms of literate, numerate and technically inclined men – for the explosion in mechanical engineering that came after 1830s with the railways and other industrial developments.

The workmen's and artisans' educational organisations, though, were not the only institutes going in the first 30 years of the 19th century. The Smeatonian Society of Engineers, so named since Smeaton's death in 1792, was one of the few organisations to be set up with the specific aim of discussing technical topics. It was, though, limited in numbers and elderly in membership, consisting of only the most senior and famous engineers of the day.

From about 1816, a group of young London-based engineers met informally to discuss engineering matters. Over the next couple of years, about eight of them, the oldest 32, the youngest just 19, decided to put their meetings on a more formal basis, and on January 2, 1818, they formed themselves into a society for young engineers up to the age of 35. Just 11 days later, at what was already the new society's third meeting, the name of the society was chosen: the Institution of Civil Engineers.

Civil engineering at this stage was so named to differentiate it from military engineering. Not until well into the 19th century would civil engineers come to be identified with the construction of static structures. The first members included young engineers employed by Telford and by Maudslay in activities ranging from construction to mechanical engineering.

The ICE's progress in its first few years was painfully slow. By 1820, with subscriptions of five shillings, membership had risen to around a dozen – and drastic action was needed. The upper age limit of 35 was scrapped and the most eminent engineer of the day, Thomas Telford, was invited to become the first president. Amazingly, as he had not even heard of the Institution, Telford accepted. Not only that, but he defined what the purpose of the Institution was to be and gave it a huge collection of books, maps and parliamentary reports. Within a matter of weeks, a fairly aimless collection of young men meeting in coffee-houses had been transformed into the basis of a modern professional Institution.

Further respectability came in 1828 with the granting of a Royal Charter. The charter contains the classic definition of the Institution: "A Society for the general advancement of Mechanical Science and more particularly for promoting the acquisition of that species of knowledge which constitutes the profession of a Civil Engineer, being the art of directing the Great Sources of Power in Nature for the use and convenience of man".

There was no suggestion at this time that the ICE should confine itself to just part of engineering: the charter, in fact, went on to list every type of engineering known to the engineers of 1828. Across the 1830s, membership grew steadily, and the Institution established itself in Great George Street, Westminster, next door to the Stephensons, in 1839.

But all was not entirely smooth. The ICE was only partly successful in attracting applications for membership from the engineers engaged in railway construction and operation, and the top posts in the Institution went unfailingly to "resident" members based in London, not to "corresponding" members from the provinces. To give it its due, the ICE was aware of these failings and attempted to encourage a wider geographical spread. The practice of holding weekly meetings, though, made the participation of busy engineers from the provinces difficult, even in the era of railways. And there were lingering doubts about the commitment of the Institution to engineers in manufacturing, as distinct from project-based branches of engineering.

What the ICE had done, though, was to present a pattern for other professions, or other parts of the engineering profession, to emulate. In 1834, the architects set up their own Institution. All across Britain, in professions far removed from engineering and among groups tied together by geography, or common interests, or by not very much at all, similar societies and organisations were being formed.

The formation, in 1847, of the Institution of Mechanical Engineers reflects the growth in mechanical engineering, particularly in railways, and the mechanisation of all kinds of manufacturing industry. It indicates a recognition that mechanical engineering represented a distinct and discrete subject, worthy of study and discussion. It suggests, without criticism, that the existing organisations such as the Institution of Civil Engineers were somehow not quite fitting the bill. It fits very much the early Victorian pattern of disseminating knowledge and information through societies and associations.

Above all, though, it was an expression of confidence in the progress of mechanical engineering and in the contribution that mechanical engineering was already making to the wider progress of mankind.

2 The workshop of the World

1847 to 1877

George Stephenson, the father of the railways and the first president of the Institution of Mechanical Engineers, was 66 years old in 1847. But most of the engineers who were the first members of the Institution in January 1847 were young men in what was very much a young profession.

James McConnell, the chairman of most of the first meetings, and Archibald Slate, the honorary secretary, were both 32 in 1847; Charles Beyer was a couple of years older. Of the four signatories to the original notice advertising the formation of the Institution, Edward Humphrys, at 38, was the oldest.

The exponential growth in all aspects of engineering was the prime reason why so many young men were involved in the early days of the Institution. There is a strong suspicion that never has engineering been quite so much fun, or quite such an attractive career. The buccaneering age of engineering lasted for perhaps no more than 30 years, from about 1830 to 1860, but it changed the face of Britain and the structure of the British economy and society. It was the work largely of young engineers with few inhibitions or constraints, applying the new technology of mechanical engineering to transport, energy production and manufacturing.

The most visible sign of the phenomenal growth in engineering was the development of the railway network. By the end of 1846, more than 3,000 miles of railway lines had been opened for goods and passenger traffic. Most of the main cities and towns in England

The epitome of mechanical engineering when the IMechE was formed: Cabry's 2-4-0 passenger locomotive of 1846/47 for the York, Newcastle and Berwick Railway, built by Wilson at Leeds and drawn by David Joy

LOCOMOTIVE PASSENGER ENGINE BUILT IN 1846-7 BY E. B. WILSON & Co. RAILWAY FOUNDRY, LEEDS.
FOR THE YORK, NEWCASTLE & BERWICK RAILWAY. THOMAS CABRY, ENGINEER.

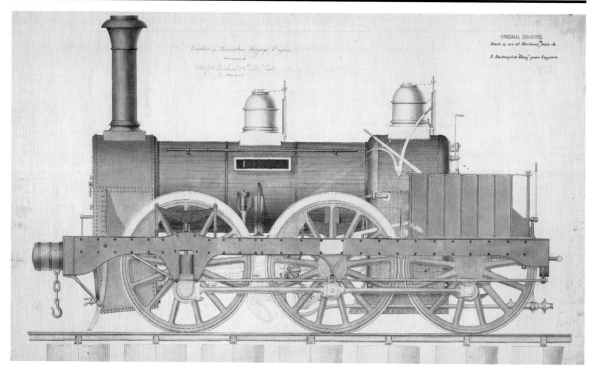

Gray's six-coupled engine of 1843/44, drawn by the engineer himself. John Gray worked with George Stephenson on the Liverpool & Manchester Railway, and later became locomotive superintendent of the Hull & Selby Railway

were linked to each other by rail, which was why the Institution of Mechanical Engineers was able to be formed as an organisation based in the provinces: Birmingham, almost as far from the sea as anywhere in Britain, was newly accessible from the north, the north-east, the south-west and from London, by rail.

The railway mileage completed by the end of 1846 was eight times that of 10 years before, but was still only a third of the total mileage that had received parliamentary approval. This was the period of the fastest growth: the total route mileage of British railways more than doubled again, to 6,500 miles, in the four years to the end of 1850. By this date, railways built, in part, to carry coal and iron ore were themselves the biggest single consumer of both: a further example of the self-sustaining circularities of the industrial revolution.

Gray's valve gear, drawn in 1846, which he patented. The gear led to Gray's downfall: he sued the London & North Western Railway for patent infringement, but lost, and died in poverty soon afterwards

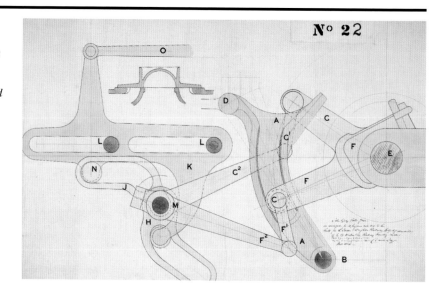

But the bald mileage statistics and the tonnages of coal and iron ore of the age of railway development do not do justice to the frenzy of excitement and activity the railways generated. In the late 18th century and at the start of the 19th century, though truncated by the national involvement in the Napoleonic wars, there had been a hectic round of meetings, subscriptions and parliamentary bills to promote canals. The canal mania was, though, nothing to the railway mania which gripped Britain from the opening of Stephenson's Liverpool & Manchester Railway in 1830 onwards and which subsided finally only at the end of the century when the long-drawn-out folly of the Great Central's London extension brought home the real economics of railway construction.

In fact, of course, the economics of railway construction and operation had been doubtful from the beginning. The reason so many of the lines planned in the great railway mania years of the 1840s were never built was that the realities of their business prospects were rumbled before too much was spent. It is arguable that cancellation was a mistake in many cases, with prospects often based on the false premise that freight was the money-earner and that passenger traffic would be minimal.

But then those railway lines that were constructed lost fortunes almost as often as they made them. The 1840s were the era of George Hudson, the railway king, whose spectacular success in railway speculation ended, to the satisfaction of later Victorian moralists, in even more spectacular failure.

One important reason why the railways were from the outset often a doubtful business proposition was the enthusiastic involvement of the engineers. It has been estimated that Britain's railways cost, on average, four times as much as those in continental Europe and in the United States. Some of that extra cost was caused by difficult terrain: very few railways in Britain were a simple matter of laying track down across existing land, as the railroad builders of America sometimes managed. Some of it, also, was the result of the need to pioneer techniques and technologies that had not been used before.

The gathering at the foot of the Lickey incline in 1846 which led to the formation of the IMechE is an example of the degree to which the railways were, in their early days, a testbed for new ideas and for experiments. Engineers were not sure, until they tried it, what degree of slope they could get away with in designing new lines. Contemporary reports noted the difference in rail construction techniques between Robert Stephenson, who tended to drive his lines through any obstacles, and Joseph Locke, who preferred generally to stick to the old canal builders' rule of following contour lines where possible. At this stage, of course, there was little in the way of engineering science to tell them what was feasible, and the empirical method was well engrained as the way to make progress through innovation.

In fact, a lot of the extra cost in Britain's railways appears to have been down to the way that the engineers, especially the civil engineers involved in construction, acted as if wholly unconstrained by cost. The result is a railway legacy of graceful structures, architecturally distinguished buildings and lines that, in many opinions, enhance rather than intrude into landscapes. But the legacy was often bought at a price.

Extravagance in the mechanical engineering side of railways was probably less commonplace and perhaps more excusable, in that developments in locomotives and rolling stock were consistently pushing forward the boundaries of mechanical engineering knowledge. In the 20 years after the Rocket, the tractive power of steam locomotives increased many times.

The railways were not the only industry in which mechanical engineering was making giant strides forward in the years around 1847. The world of shipbuilding was similarly being revolutionised by the steam engine and by iron; there were important developments in agricultural machinery, armaments and machine tools. Mechanisation, which had started from manufacturing industries such as textiles, was beginning to break out of the factory to affect a wider range of products and devices and a wider span of everyday life. Other forms of engineering were starting to deliver services such as water supply, sewage disposal and gas supply to domestic and industrial premises.

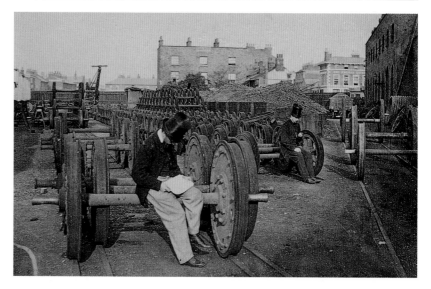

The stockyard of the Great Eastern Railway at Stratford in around 1860. The figure in the foreground reading is believed to be William Henry Maw, IMechE president in 1901 and long-time editor of Engineering, *who started his career at the GER*

The railways, though, were the pace-setters, and not just in the speed of their development. They were also important, for instance, in the largely unseen transition within engineering from the craft-based workshop of Boulton and Maudslay to the factory system where series production of identical products was organised around specific skills.

The steam locomotive was one of the first large mechanical engineering products to be made in bulk. Before the railways, virtually all steam engines had been built to order and were one-offs. The new railway operating companies wanted their engines in greater numbers than had ever been built before, and they also wanted reliable performance and at least a degree of standardisation. In the early days, the demand ran well ahead of the supply, and "engineering" companies that had little long-term interest in steam engines won orders or found themselves subcontracted to supply locomotives.

The variability of performance and quality that resulted and the degree to which railway companies were not in control of their costs

Gas lighting was one of a number of services which started to be supplied to towns and to industrial premises from the middle of the 19th century onwards

were reasons why the companies moved quickly to set up their own workshops for newbuild and repair. Again, in this respect the railways were setting a trend: across the 1850s, in the important railways, many of them formed by bringing together local lines to create national routes, there are the unmistakable signs of modern-day business issues and values being asserted.

The recognisably modern manufacturing company also dates from this time, and the corporate values of frugality, solidity and responsibility to shareholders start to take precedence over the adventure and innovation of the paternalist engineer-entrepreneurs of the early part of the 19th century.

One of the IMechE's founding fathers appears, eventually, to have been a casualty of the more business-like approach.

James McConnell seems, from the early proceedings, to have been not just the host whose hospitality led to the formation of the Institution, but also its most colourful personality. The transcriptions of early discussions indicate that McConnell was usually in the thick of the debates that followed the presentation of papers; for the first half dozen years of the Institution's history, he was the regular deputy as chairman of Institution meetings for the two Stephensons who were the first presidents.

James McConnell's 2-2-2 locomotives for the London & North Western Railway's southern section were nicknamed Bloomers because they exposed parts normally unseen, as the fashion underwear of the 1850s did

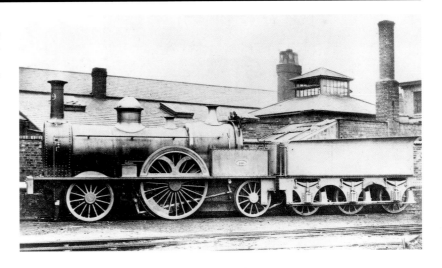

The Large Bloomers of 1861 were the last locomotives built by McConnell before he was removed from the LNWR by the parsimonious management regime of Richard Moon

William Fairbairn's Manchester factory built several locomotives for McConnell

McConnell's career is typical of the younger engineers who formed the Institution. A Scotsman who was born in Ireland, where his father had bought a foundry, he was apprenticed for eight years from the age of 13 to a Glasgow engineering firm, then moved to Liverpool. There, in the first railway boom of the late 1830s, he was involved in a company whose main activity was shipbuilding but which also built some steam locomotives for the London & Birmingham Railway. The company's head was the locomotive innovator Edward Bury, who seems to have become McConnell's patron.

McConnell then had at least two other jobs in workshops in Liverpool and Manchester before being appointed locomotive super-intendent of the Birmingham & Gloucester Railway at the age of 27 in 1842, apparently on the recommendation of Bury.

McConnell's knowledge of railway practice was fairly limited, but his experience was little different from many of the other engineers who played prominent parts in railway development. McConnell seems, in any case, to have made something of a name for himself already with his work in Liverpool for the London & Birmingham, putting into practice his own ideas on balancing the moving parts of

McConnell is credited with at least part of the design of this steam carriage by a Mr Rickett of the Castle Foundry, Buckingham, shown here by the Illustrated London News *in 1860 on its way to be presented to the Queen and the Prince Consort*

STEAM-CARRIAGE TO RUN ON COMMON ROADS, DESIGNED BY MR. RICKETT, OF THE CASTLE FOUNDRY, BUCKINGHAM.

the locomotive – a practice thought unnecessary by other engineers, but which McConnell believed would prevent uneven wear on tyres and accidents caused by locomotives jumping the rails. Engine balancing was also the subject of his only paper presented to the Institution of Mechanical Engineers in 1848.

Unlike some engineers who ended up in positions of influence in the early days of the railways, McConnell had a real flair for locomotive design: his Great Britain engine for banking heavy trains up the Lickey incline was probably the first saddle-tank locomotive, and he pioneered a double furnace engine and the use of coal for firing instead of the more usual, and more expensive, coke.

In 1847 in a distinct promotion, he took up the job of locomotive superintendent for southern division of the London & North Western Railway, the company formed by the amalgamation of the London & Birmingham and the Grand Junction Railway which had already subsumed the Liverpool & Manchester. McConnell was in charge of the Wolverton works, in what is now Milton Keynes, and from there he produced the famous Bloomer 2-2-2 engines, so called because the high boiler exposed parts normally unseen, in the manner of the bloomers which were the fashion underwear of the 1850s.

The LNWR's arrangement of having a separate locomotive superintendent, with separate locomotive designs, for each of its divisions was always likely to create tensions and inefficiencies. McConnell's counterpart at the main northern works at Crewe was Francis Trevithick, son of a distinguished father but not a great engineer himself. At Longsight in Manchester, John Ramsbottom, also a founder member of the IMechE, had a less direct role in locomotive design, but was increasingly influential in the small and simple – but remarkably powerful – engines that emerged from Crewe. It was an uneasy triumvirate.

Towards the end of the 1850s, the threat that the ambitious Ramsbottom might leave for a chief engineer's job elsewhere or into consultancy appears to have brought the issue of the three engineers to a head. The general manager of the LNWR was the determinedly cost-cutting Richard Moon, whose name became a byword for

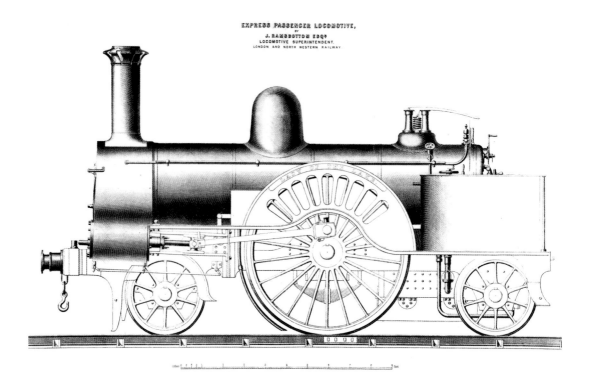

EXPRESS PASSENGER LOCOMOTIVE,
BY
J. RAMSBOTTOM ESQ?
LOCOMOTIVE SUPERINTENDENT.
LONDON AND NORTH WESTERN RAILWAY

John Ramsbottom's Lady of the Lake class was his 2-2-2 express locomotive from the LNWR's main works at Crewe, a more conventional design than McConnell's Wolverton efforts

Victorian parsimony. Moon manoeuvred Trevithick out and Ramsbottom into the top engineering job at Crewe; he also started a campaign against the alleged extravagances of McConnell at Wolverton, where the flamboyant engines in brilliant scarlet livery were in stark contrast to the soberly efficient workhorses coming out of Crewe.

The campaign against McConnell took some years to work. Quite why it should have affected his involvement in the Institution is not clear, but by 1860, McConnell was playing very little part in the affairs of the IMechE and the following year he is no longer in the list of members. In 1862, at the age of 47, McConnell "retired" from the LNWR, spending the last 21 years of his life in some obscurity as a "consultant" in London. His death in 1883 went unrecorded in the IMechE Proceedings, an omission that sits uncomfortably alongside the punctillious obituarising of engineers who contributed far less to the Institution.

McConnell's conviviality seems to have been an important ingredient in the early years of the IMechE. He was also generous, in the style of the paternalistic owner-employers, to his workforce at Wolverton, being responsible for the building of schools, churches, a savings bank and a mechanics' institute in the town. When he left, the workforce clubbed together to buy him a piece of high Victoriana, described in a contemporary newspaper report as "an elegant and massive silver epergné or candelabrum with cut-glass centre dish for flowers, and six branches supporting cut-glass dishes, which may be removed for the purpose of holding candles; the stem represents the vine with leaves and grapes, beautifully modelled and arranged". As well as the gift, there was "an appropriate address bearing 1,966 signatures".

Conviviality as well as serious purpose seems to have been the flavour of the Institution of Mechanical Engineers from the beginning. The rules adopted at the first meeting on January 27, 1847, specified that four meetings were to be held each year on the fourth Wednesdays of January, April, July and October, with papers to be read by members at each of them. The first meeting of the year was to be the annual general meeting at which the officers for the next year would be elected.

The 1847 meetings were all held at the Queen's Hotel and were rather irregularly spaced. On May 17, the first paper was presented by William Buckle of Boulton & Watt, who was a member of the Institution council, on "A series of experiments relative to the Fan Blast". The paper was primarily about the design of centrifugal fans, but Buckle also discussed the possibility for using fans in blastfurnaces, a development that did not become reality for 50 years.

Edward A Cowper – the middle initial being used to distinguish him from his father, the professor of engineering at King's College London – was, at 28, the youngest member of the Institution council. He gave the next paper on suspension bridges to a meeting in October. The subject appears an odd choice for the new Institution, and there is no record of any discussion. Cowper was, at this stage in his career, much concerned with structural steelwork: he was working at the Birmingham office of Fox Henderson, the foundry and structural engineer, and was responsible there for the roof sections of the Crystal Palace, built for the Great Exhibition of 1851, and for the roof of the New Street station in Birmingham that replaced Curzon Street. He later invented the Cowper stove, the regenerative heater for the air supply to blastfurnaces, and described it in a paper to the Institution in 1860; alone of the first council of the IMechE, he went on to become president.

The record of the first couple of years of the IMechE fuels the suspicion that it was the efforts of a very small number of engineers that established the Institution and then sustained it through the critical opening years. The early papers were almost universally given by members of the Institution council, and reports of the discussions that followed in the proceedings see the same names cropping up many times. The rule that all members should produce one paper a year seems never to have been enforced, and was quietly dropped in 1855.

In November 1847, another council member, Charles Beyer, the man who with McConnell can be credited with the original idea for the Institution, gave a paper that was to prove quite a rarity in the long history of the Institution. Although the railways and their development were central to the formation of the Institution, and many of the engineers who were early members were employed on the railways, very few papers in the early years dealt specifically with individual locomotives. The IMechE seems to have been concerned from the beginning to keep discussion as wide-ranging as possible.

In his paper, Beyer described the "luggage engine" Atlas that he had built at Sharp Brothers in Manchester for the Manchester, Sheffield & Lincolnshire Railway – the line of which Richard Peacock, another of the founding fathers and later Beyer's business partner in the locomotive building company Beyer Peacock, was chief engineer.

Charles F Beyer, the German emigré who was one of the moving spirits behind the formation of the IMechE. The painter of this portrait, held at the institution, has not been identified

The luggage engine, more commonly known as a goods locomotive, was, by 1840s standards, big and powerful, but scarcely innovative.

Beyer's own career reads like one of the more implausible Victorian morality stories. Born the son of poor handloom weavers in Saxony, Karl Beyer was taught drawing and mathematics by a family friend, but when he was 12 he was apprenticed to a weaver, and that looked likely to be his lot. But a local doctor saw one of the boy's drawings, and persuaded Beyer's father to send him to the new polytechnic in Dresden. The family's poverty meant Beyer had to rely on charity to finish his studies, and he then worked in a textile machine company at Chemnitz.

In 1834, at the age of 21, Beyer was commissioned by the Kingdom of Saxony to visit Manchester and other northern towns in England and report on the textile spinning machinery used there. He reported back, and two joint stock companies were set up in Chemnitz and Dresden to spin cotton. Beyer was offered the job of managing either of them, but refused. By the end of 1834, he was back in Manchester where he managed, after some difficulty, to obtain work as a draughtsman with Sharp Roberts, the biggest textile machinery manufacturer and a company rapidly moving into steam locomotive building. Within a few years, he was head of the locomotive side of the business, and when the architect of the Manchester tradition of toolmaking, Richard Roberts, retired in 1843, he became the engineering chief.

Beyer was not the only German engineer to make the journey to Britain in search of greater engineering scope: Henry Dübs, an engineer at Tayleur's Vulcan foundry in Warrington, was like Beyer a founder member of the IMechE; and William Siemens would, within a few years, outshine both of them as an inventor, as IMechE president and as the recipient of a knighthood. In each case, the attraction appears to have been the greater degree of engineering development in Britain than in the German states and the relative openness of British society.

Beyer was one of a number of Manchester-based engineers who appears to have made time to take an active part in IMechE affairs; another council member from Manchester, Benjamin Fothergill, from the textile machinery side of the Sharp Brothers business, gave the paper that accompanied the January 1848 annual general meeting, speaking on the "Multifarious Perforating Machine" and other textile developments.

The annual meeting was able to report growing membership and a healthy bank balance of £218 6s 5d at the Birmingham branch of the Birmingham & Midland Bank. With subscriptions at £5 for the first year and £3 thereafter, it is hard not to conclude that the founding fathers must have done a first-class selling job in persuading engineers – 162 of them by the end of 1848 – to part with their money.

A further indication of skill in money-raising came at the April 1848 meeting, when 16 distinguished men who had so far failed to join were invited to become "honorary members". "Honorary", in this instance, did not infer "gratis": the 16 duly paid their subscriptions like everyone else. Some of them were, even by the fairly open qualification procedures of the early IMechE, dubiously qualified for membership. Henry Heane, for instance, was secretary to the Shropshire Union Railways Company, and William Overend was a London barrister. The Institution appears to have learned quickly that it helps any new organisation to have the "right" people on board.

Institutional hindsight has rather endowed the summer 1848 meeting of the IMechE with a glow of nostalgia. The meeting, like the April one before, was held not at the Queen's Hotel but at the Philosophical Institution in Cannon Street, Birmingham, and was presided over by George Stephenson.

If Stephenson was initially confused about the nature of the IMechE when he accepted the presidency, the members had had little to complain about in his commitment since. He had missed barely a meeting and, from the reports of the discussions in the Proceedings, appears to have acted as a wise, fair and involved chairman. Doubtless, though, the younger engineers who had invited Stephenson to be their president were hoping for even more, and on July 26, 1848, he obliged.

The Proceedings reported: "The President said that the first Paper to be read, according to the programme, would be his own, 'On the Fallacies of the Rotary Engine'."

To call what followed a Paper was probably a degree of exaggeration. Stephenson's contribution was barely 200 words and just four paragraphs, and most of the words refer to a single diagram that he had drawn. His target was an engine, a model of which had been

George Stephenson's paper on
the self-acting railway brake –
note the normal Victorian
spelling

George Stephenson's paper on
the self-acting railway brake –
note the normal Victorian
spelling

brought to the same meeting by a Mr Onion. Stephenson had had experience of rotary engines before: he claimed he had been stranded at sea on a ship powered by such an engine and that it had cost him £40 to be towed back to Yarmouth. His dismissal of the engine is terse and to the point: about 50 patents had been taken out on rotary engines, he said, and every one of them had failed. "No man who ever lived could improve on the lever principle, as there is no power but in the lever."

Onion, who was not a member of the Institution, appears to have tried to defend himself, saying that his engine had been working for some weeks at the Midland Railway's Derby station. But Stephenson was having none of it: one trial was not a proof, he said. The engine might work once and then fail when it was tried again.

Joseph Miller, with McConnell and Beyer one of the three vice-presidents and a London marine engine builder, attempted to defuse the row by saying that if rotary engines could be made to work, they would have advantages in terms of compactness. But Stephenson thought there was no merit in Onion's engine at all.

The discussion moved on to other matters, with William Buckle presenting a paper on "a machine for preparing bone manure". The George Stephenson paper, though, was lent an aura of sanctity by what happened next. Just over two weeks after the IMechE meeting, and after a very short illness, Stephenson died. Though there are many records of his writings in letters and company prospectuses, and other written contributions to Institution debates, notably on railway braking systems, the paper to the July meeting in Birmingham is the only one Stephenson delivered in person to a learned society in his long career.

Stephenson's death at 67 was a shock, but the IMechE rose to the occasion. If there was one thing that the Victorians did well above all others, it was grief.

The Institution council met – unusually, in Manchester – and wrote to Stephenson's widow, his former housekeeper whom he had married as his third wife at the start of 1848. It also commissioned John Scott Russell, the Thames shipbuilder who is more closely associated with Isambard Kingdom Brunel than with the Stephensons – he was the designer and builder of the Great Eastern – to write a eulogy for delivery at the next meeting of the Institution in October.

Even by the standards of such things, Scott Russell's address was overstated. George Stephenson, he wrote, "will be known to posterity as the presiding Genius of our times". He had been responsible for the introduction of "Railways and the Locomotive Engine", which were not merely elements of social progress but had also made a material contribution to the peace that had existed since 1815.

It was not just Stephenson's inventiveness which Scott Russell wanted to praise. Engineers, he remarked, perhaps with Brunel as much as Stephenson in mind, were often difficult characters: "The continual struggle with tough, hard and refractory substances, which forms the business of the engineer, has the effect of communicating a hardness of character, an obstinacy of disposition and a rigidity of temper to men of our craft which does not add to their excellence as members of society." Not so Stephenson who was "happily for himself and the world, a man endowed with no common share of the endowments which make the intercourse of life useful to himself and delightful to his friends".

He went on: "In his latter days he was distinguished for the childlike simplicity of his character, for the transparency of his intentions, for the singleness of his purposes and for the straightforward manly honesty of his conversation and dealings. If he could hate an enemy, he never masked his antipathy by hypocrisy; but he was a warm and earnest friend."

Some of the effect of Scott Russell's encomium to George Stephenson may have been lost because the author was unable to deliver it himself. As ever, the task devolved on McConnell. The Proceedings noted that the address, "both at the close and during the time of reading, elicited expressions of admiration".

There were, though, some rather more waspish comments about Stephenson in the preamble to the formal motion expressing the regret of the Institution. Thomas Geach, the banker who was the IMechE's treasurer, proposed the motion and managed to get away with a few sideswipes about Stephenson's cantankerousness. The Proceedings reported Geach's speech in the stilted style of formal reports: "He would quite allow that his manners were sometimes rough – he would quite allow that there were peculiarities in his character which had to be considered as peculiarities. He was willing to allow that he had seen in Mr Stephenson what in other men might subject them to criticism; but when it came from Mr Stephenson, it came from a privileged person." Stephenson, Geach added, was proud of his early career, "and never lost any opportunity of expressing it".

The motion expressing regret was, of course, passed unanimously. McConnell then went on to say that the council, meeting after George Stephenson's death had made other decisions. "They resolved that the best tribute they could pay to the late Mr Stephenson's memory, and the best way in which they could testify their appreciation of his merits, besides at the same time the best selection of a future president they could make from among the eminent men of the day, would be to invite Mr Robert Stephenson to become his father's successor, as president of this Institution."

This decision seems to have been followed by something of a farcical interlude. Two of the older members of the council, William Buckle and Benjamin Fothergill, had been sent off to London "to wait on Mr Stephenson". They missed him. Quite why they missed him is not clear: McConnell refers to "an accident". It seems probable that Robert Stephenson was not, in fact, in London at the time, and it speaks volumes for the earnest self-importance of the Institution council that they should have supposed that he would be. The invitation to become president had to be sorted out later by letter.

Robert Stephenson was a very different character from his father. George Stephenson's first wife had died when Robert was a baby, and though George later remarried – twice – the father invested much of his energies and his attention in the son. Robert, he determined, should have the education that he had lacked and he sent him to the grammar school in Newcastle upon Tyne and later arranged private tuition for him in Edinburgh, where he attended lectures given by the university professors of mathematics, chemistry and geology.

Robert Stephenson, the portrait by John Lucas, who was also responsible for the portrait of George Stephenson in the possession of the IMechE

Less formal sketch of Robert Stephenson by the artist George Richmond

Robert Stephenson may not have been an inventive genius on the same scale as his father, but he was one of the great engineer-administrators. In his early 20s, while his father was working on the Liverpool & Manchester Railway, he took charge of the locomotive works George had set up on Tyneside, and set about making the steam locomotive a more reliable machine. The credit for the Rocket's success in the Rainhill trials is as much due to Robert, as to George Stephenson. In the 1830s, as one of the great railway builders, he was responsible for a construction workforce of around 10,000 men on the London & Birmingham Railway. His construction technique, to go through the obstacles rather than around them, meant some civil engineering innovation: the line has, at Kilsby, the first of the great tunnels on a main railway line. His later fame derives very much from his bridge-building exploits at home and overseas, including the high level bridge over the Tyne in his native Newcastle.

Partly because of his education, Robert Stephenson was respectable in a way that his father never could be, and perhaps never wanted to be. He had none of the disdain of George for London, and by the 1840s was operating from the capital as a consulting engineer for railway projects across Europe and beyond. His address was 24 Great George Street, Westminster, next door to the Institution of Civil Engineers. In 1847, he was also elected as Member of Parliament for Whitby.

These business and parliamentary commitments meant that Robert Stephenson was not a particularly committed president for an Institution based in Birmingham, though there is plenty of evidence that he kept in touch with the Institution and helped it gain in prestige and status. But the record of his attendance at meetings in the five years of his presidency is not good. In 1851, the IMechE held its first-ever meeting in London during the Great Exhibition, but Stephenson failed to attend that as well, though he did manage to chair the banquet held at the Freemasons Tavern in the evening. In addition, he missed all four of the regular meetings that year.

If the members never complained, Stephenson himself seems to have recognised that he was not doing the best job he could for the Institution. He chaired the October general meeting in 1852 and told the Institution that he had been instrumental in changing the rules at

The Robert Stephenson business card indicated the range of his engineering interests

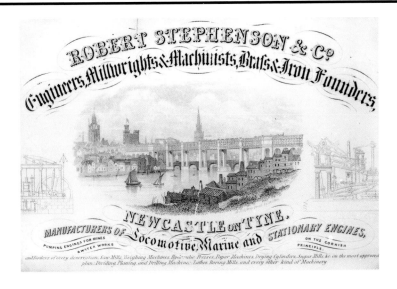

William Prime Marshall, former colleague of Robert Stephenson and secretary of the IMechE from 1849 to the move to London in 1877

the Institution of Civil Engineers to scrap the "injurious" system of life presidents. He now proposed the same for the IMechE. The Institution council had asked him to stay on for one more year, and he had agreed, but he suggested that presidents should then be elected for no more than one or two years.

In theory, by suggesting he could stand down, Stephenson was doing no more than applying the rules: a change to the regulations in 1849, following the death of George Stephenson, had made it compulsory to re-elect the president each year. In practice, of course, until Robert Stephenson indicated that the presidency ought to change, no member was likely to challenge him. It was an important point, and freeing up the presidency allowed the Institution to move forward gradually from the era of the railway pioneers and from a potentially unhealthy dynastic succession.

In Stephenson's absences, the Institution's meetings were chaired by a variety of council members, with McConnell prominent as ever. These were years of growth and consolidation for the new Institution, with membership barely increasing but the trappings of a prestige organisation being acquired. As early as the middle of 1847, it was apparent that Archibald Slate as honorary secretary could not handle all the administration needed with the membership, and a permanent secretary, Archibald Kintrea, was appointed. Kintrea lasted for little more than a year, and was replaced by William Prime Marshall, who had been Robert Stephenson's assistant and chief engineer of the Norfolk Railway.

A further step towards permanence was taken in 1850. The Institution acquired its first premises on Newhall Street, Birmingham, with a meeting room, a library and accommodation for Marshall. Robert Stephenson was unable to make the opening of the new offices, as he was in North Wales for the floating out of the last tube section of his Britannia Bridge across the Menai Strait. He sent a £100 cheque instead.

Honorary membership had been introduced in 1848, and two years later a new class, graduate membership, was added. The aim was to attract younger engineers who were not yet in charge of engineering businesses or who had not had five years experience, and McConnell expressed the hope that graduate numbers would soon rival the numbers of full members. It was a vain hope for the time. With few engineering courses yet providing graduates and with no real incentive to take up this class of membership – the age limit and the subscriptions were exactly the same as for full membership – the number of graduate members reached just four in 1852, and then fell away again. In 1866, the rules were changed again so that full membership was available only to those aged 24 and over, while the age limit for graduate members was reduced to 18. Graduate membership then became the only route to full membership for those younger than 24.

If the official policy of the Institution in its first five years was to open the doors as wide as possible to attract new young members, there is little doubt there was an unofficial policy geared to persuading senior industrial figures to join. Apart from the Stephensons and John Scott Russell, the list of founder members lacked heavyweight figures, though some were to achieve considerable fame later in their careers.

The rule change of 1849 which instituted the annual election for the president also increased the number of council members from five to 15 and three years later a further change increased the number of vice-presidents from three to six. This was part of a process of broadening the Institution's appeal and had the beneficial effect of creating senior positions for senior engineers. Among the new names which came on to the council of the Institution for the first time as a result of these changes was John Ramsbottom, McConnell's London & North Western Railway colleague, and later his rival.

Perhaps more significant, though, was the introduction of two heavyweight industrial figures to the upper echelons of the Institution. William Fairbairn and Joseph Whitworth were perhaps the two most famous engineers in manufacturing after Robert Stephenson. Both were Manchester-based and had joined the Institution in the second wave of members at the end of 1847. Whitworth was a fitful participant in Institution affairs in the first few years, but Fairbairn had taken no active part whatsoever.

Of the two, Fairbairn is probably less known to today's engineers, since his innovative credits are a little more dubious. His range of activities, though, was huge and indicates the breadth of interest of the engineer-entrepreneurs of the Industrial Revolution.

Fairbairn's fortune, which was large, was derived from using iron in water wheels and from the machinery used in the textile industry; he had also been a partner of Robert Stephenson in a not entirely successful shipbuilding venture on the Thames in London. He was a pioneer in structural ironwork and bridges, and was again Robert Stephenson's partner on the Britannia Bridge. In terms of innovation, his company appears to have been instrumental in the development of the Lancashire boiler from the Cornish boiler made originally by Trevithick and in developing the steam rivetter.

Where Fairbairn was particularly innovative and influential was in developing Britain's trade in mechanical engineering. In 1839, a speaker at an Anti-Corn Law League meeting in Bolton complained that Fairbairn's company "did little for England" and that his workforce was "constantly employed in making machinery to be sent abroad". This was not true, but Fairbairn did have a huge export trade, and visited countries including Turkey, Sweden and Russia himself.

Fairbairn's accession to the IMechE presidency is odd. In 1853, having taken no part in the Institution at all before, he was elected as a vice-president, apparently on the recommendation of Robert Stephenson. There seems little doubt that Fairbairn, at 64 one of the oldest members of the Institution, was Stephenson's chosen successor as president, and he was duly "elected", unopposed, at the annual meeting at the start of 1854. He had, in fact, also chaired the last meeting of 1853 in Stephenson's customary absence: this was the first general meeting of the Institution to be held in Manchester. After serving his two years as president, Fairbairn remained as a vice-president for a further three years, then left the Institution entirely, not even remaining as a member, though he seems to have maintained his membership, and active participation, in other organisations.

What Fairbairn brought the IMechE was industrial muscle and that, in the 1850s, conferred on the Institution a respectability and gravitas that was essential to its success. His Manchester and London connections helped establish the Institution's credentials as a national body. Fairbairn also started the idea of a presidential address, though only Joseph Whitworth followed the practice until, in the mid 1870s, Frederick Bramwell revived the concept.

To Fairbairn's status in manufacturing industry, Whitworth added innovative lustre. Whitworth's renown as an industrialist was much enhanced by the Great Exhibition of 1851, where his display of machine tools had been one of the features.

Joseph Whitworth, who brought the institution into the forefront of industrial debate

The IMechE was not involved officially in the organisation of the Great Exhibition. But the exhibition was a material help in establishing the Institution as a national and influential body; several important members, including the president Robert Stephenson, were involved in drawing up the plans and judging the exhibits, and the Institution showed a happy knack of picking the right issue on which to concentrate.

The idea for a national exhibition of British manufacturing and commercial prowess had stemmed from an event on similar lines in Paris in 1845. The British plan was initially pursued through the Royal Society of Arts, which held a series of local exhibitions from 1847, attracting audiences of up to 20,000 people. The notion of a national exhibition was then taken up by Henry Cole, assistant

The accuracy achieved by Joseph Whitworth's machine tools was one of the sensations of the 1851 Great Exhibition

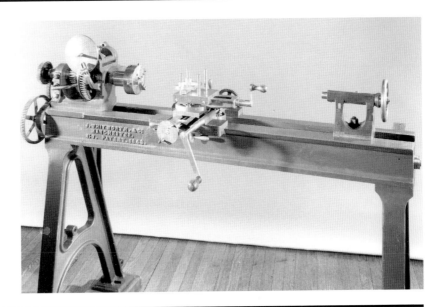

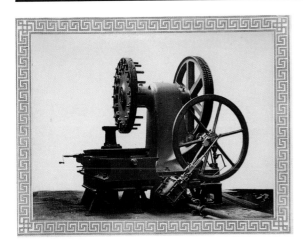

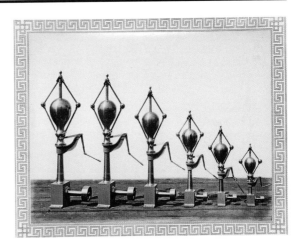

Whitworth machinery: different sized governors and an armour-plate milling machine from the 1850s

keeper of the Public Record Office, and it was Cole who persuaded Prince Albert, the Queen's consort, to lend his backing.

Prince Albert put forward a plan for a national exhibition to be under the control of a Royal Commission, but the government declined to be involved. Despite this, a group of manufacturers in Birmingham then went ahead with their own version of a national exhibition in 1849, and 100,000 visitors attended. Convinced by the Birmingham success that the idea was sound, Cole urged Prince Albert to try again: this time, the government was persuaded, and offered Somerset House on the Strand as a venue.

But by now both Cole and the prince were intent on something more ambitious than the limited accommodation in Somerset House could provide. They proposed the construction of a huge temporary hall in Hyde Park and the expansion of the event into a national and international exhibition to demonstrate global trade and the interdependency of nations. There would be an emphasis on technology and manufactured goods, but the exhibition was also to cover the arts, handicrafts and agricultural products from around the world.

The proposal to use Hyde Park aroused considerable opposition. In the House of Commons, the prototype for all caricatures of reactionary pomposity, Colonel Charles Sibthorp, railed against the idea. "It is the greatest trash, the greatest fraud and the greatest imposition ever attempted to be palmed upon the people of this country," he said. "I would advise persons residing near the Park to keep a strong look-out over their silver forks and spoons, and their servant maids."

But the exhibition had generated an unstoppable momentum. Hyde Park was approved as the venue; Prince Albert was granted his Royal Commission; Robert Stephenson was appointed to head the organising committee; and, in perhaps the strangest move of all, Joseph Paxton, the head gardener at the Duke of Devonshire's country estate, Chatsworth, who had been the friend of George Stephenson, was commissioned to create a massive greenhouse-style structure from glass and iron – the Crystal Palace.

IMechE members contributed in large measure to the Crystal Palace itself. Apart from Cowper's design work on the roof, the overall detailed design for the whole of the building was undertaken by Fox Henderson & Co, both the principals of which had been founding members of the Institution. As a codicil, the first national meeting of

The 1851 Great Exhibition: the roof of Paxton's Crystal Palace was designed by future IMechE president Edward A Cowper

1937 was a lecture by Major C C B Morris, chief officer of the London Fire Brigade, on fire fighting. It was accompanied by a cine film of the destruction of the Crystal Palace by fire the previous year. So the Institution can claim to have been there at the great building's beginning and at its end.

The Great Exhibition opened its doors to the public on May 1, 1851, and was an instant success. In four-and-a-half months, more than six million visitors passed through the exhibition halls. The exhibition was reviewed enthusiastically worldwide. For one British writer, it was "the triumphant cock crow of the country that boasted itself the workshop of the world" and there was little doubt that the mechanical engineers and other manufacturing exhibitors were the stars of the show. British exhibitors won most of the prizes for manufactured goods, though there was a lot of admiration for American mass production ideas, particularly those shown by the Colt gun company.

Henry Cole wrote in the catalogue that the exhibits in the mechanical engineering section "form the most important series exhibited". The exhibitor names are themselves a catalogue of mechanical engineering fame. W Fairbairn & Sons showed a riveting machine for wrought iron boilers. James Nasmyth had a steam hammer able to

Inside the Crystal Palace: the south transept in 1851

impart forces large enough to shape metal or small enough to break an eggshell. Charles Beyer's company Sharp Brothers had a slotting machine and a self-acting planing machine. There were marine engines in the displays of both the Boulton & Watt and the Maudslay Son & Field companies.

The machinery sections of the Great Exhibition were admired for their gleaming metallic surfaces, and the surface finish of the machined components was much remarked on. But the exhibition had a wider and deeper effect on engineering industry and the public perception of mechanical engineering: it instilled the concept of precision.

One of the features of the exhibition was a display of precision American door locks and this, apparently by pure chance, was one of the subjects for discussion at the hastily arranged Institution of

Machinery on display at the Great Exhibition included food processing equipment: the catalogue describes this as a "vacuum sugar apparatus" by Heckmann

Agricultural machinery was displayed outside the Crystal Palace in Hyde Park

This Great Exhibition display is recorded as "turbine by Fromont"

Mechanical Engineers meeting on June 30 which was the Institution's first-ever London meeting and the first outside Birmingham.

Locks, perhaps for the only time in history, were controversial in the Great Exhibition. An American locksmith named Hobbs had excited widespread public comment by successfully picking a lock of British manufacture that had been considered unpickable for many years before. Worse still, Hobbs picked the lock in just 17 minutes. The British manufacturer suspected foul play and at the IMechE meeting challenged Hobbs to pick two locks affixed to a door, the locks to be inspected before by the chairman of the meeting, James McConnell. Hobbs replied that he had come to Britain not to pick locks but to sell them. He said he was willing to pick any lock, but he then upped the stakes by challenging the meeting to take one of his American locks apart, put it back together and then pick it – he would, he said, pay £1,000 to anyone who could do that. At this point McConnell stepped in to defuse the row.

The locks incident had two effects: it propelled the precision aspects of engineering to the fore at the Great Exhibition alongside the more visible factors of metalworking and power, and it provided the IMechE with welcome publicity.

The work of Joseph Whitworth had very similar effects over a rather longer timescale. Almost unimaginably for modern engineers used to public indifference to their products, Whitworth's display at the Great Exhibition attracted vast crowds. His Manchester factory showed a huge array of machine tools: self-acting foot lathe and a self-acting duplex lathe, a planing machine, drilling and boring machines, a slotting and shaping machine, a punching and shearing machine, and a "wheelcutting and dividing" machine, all of them attracting public attention and the acclaim of the competition judges

Whitworth memorabilia: an invitation to the 1857 summer meeting in Manchester and the later Whitworth Gold Medal

for their precision. But the undeniable star was Whitworth's micrometer, capable of measuring to a millionth of an inch.

Whitworth had made the quest for accuracy and standardisation very much his own in the 20 years since he moved from London to set up on his own as a toolmaker in Manchester. Exactly how many of Whitworth's ideas were derived from his earlier employment with Henry Maudslay is not clear, since Maudslay was fairly secretive about such matters. The basic concept of deriving accurate measurement from a screw method – end measurement – rather than from marked lines was Maudslay's, rather than Whitworth's, but Whitworth improved the accuracy of the micrometer a hundred-fold. The system of screw threads similarly had its origins with Maudslay but was expanded upon and then made public by Whitworth in 1841: it made his reputation throughout industry and beyond.

But if Whitworth's screw thread system was widely accepted and the accuracy of his machine tools widely admired, there was still one area where his influence had so far failed to make the impact he wanted. The question of accuracy of measurement and standards of length had managed to make it on to the political agenda in the early 1840s when the government of Sir Robert Peel, who was a friend of Whitworth's, set up a Royal Commission to determine how standards for length should be set. The Royal Commission was chaired by the Astronomer-Royal, Sir George Airy, a man of considerable scientific reputation and pomposity, whose every dealing with engineers seems to have been somehow less than satisfactory.

Airy and his fellow commissioners took evidence from Whitworth and then managed to spend 11 years discussing and debating their conclusions before the coming of the Crimean War in 1854 forced them into a conclusion. With a brief to come up with methods to define standard lengths, rather than to promote the interchangeability of parts that was being demanded by engineers, Airy's commission ended up by satisfying no-one, least of all Whitworth. It proposed a new Weights and Measures Bill with standard lengths based on the length of five different bars – all of which were demonstrably not the same length. Incredibly, the Bill enshrining this as the British standard passed through the House of Commons, but the House of Lords proved more sensible, and set up a select committee to consider the question.

By the time Whitworth became president of the Institution of Mechanical Engineers in 1856, he had already given more evidence to the select committee and was waiting for its deliberations. He had also spent a large proportion of the previous year working for the government on the re-equipment of the Royal Small Arms Factory at Enfield to precision manufacturing standards and, while there, he had made suggestions that radically improved the performance of the basic army rifle – and changed the course of his own manufacturing career away from toolmaking and into armaments.

So he was a man with a substantial fortune, substantial influence – and, because of the way his advice to the Airy commission had failed to be acted upon for a dozen years, a substantial axe to grind. The result was that Whitworth's presidential address, delivered at the Institution's first-ever summer meeting away from Birmingham, and the follow-up speech he delivered to the Institution at Manchester the following year, have become defining moments in the history of engineering and among the most important papers ever delivered to the IMechE.

Fittingly for a presidential address, Whitworth's first speech to the Glasgow summer meeting of September 1856 was wide-ranging and, at times, light-hearted. He bravely paraphrased one of Dr Johnson's less than complimentary aphorisms about Scotsmen and their relations with England, but also praised Glasgow's contribution to engineering and "the mechanical arts". He advocated the introduction of income tax rather than specific taxes on individual items. But more importantly for engineers, he went back to basics on "the two great elements in constructive mechanics – namely, a true plane and the power of measurement".

The true plane, obtained by making three planes, preferably from cast iron, was the standard for all references: "All excellence in workmanship depends upon it." To illustrate measurement, he had his micrometer with him and said that in end-measurement devices of this type "we have all the accuracy we can desire".

He said: "We find in practice in the workshop that it is easier to work to the ten-thousandth of an inch from standards of end measure, than to the one-hundredth of an inch from the lines on a two-foot rule."

Whitworth was, of course, right. But he was also something of a fanatic. The corollary to having correct measurement, he thought, was to have standardisation of sizes and he urged railway engineers to "consider and determine not only the fewest possible number of sizes of engines and carriages that will suffice, but also how every single piece may have strictly defined dimensions".

Nor was he going to stop at engineers in his drive for standardisation: "Suppose that the principal windows and doors of our houses were made only of three or four different sizes. Then we should have a manufactory start up for making doors, without reference to any particular house or builder. They would be kept in stock and made with the best machinery and contrivances for that particular branch; consequently, we should have better doors and windows at the least possible cost."

Whitworth returned to his themes with a more conventionally delivered paper to the Manchester summer meeting the following year, 1857. The paper is entitled "On a standard decimal measure of length for mechanical engineering work, &c" and it contains a ringing condemnation of the inaccuracies produced in engineering by the old practice of using fractions in measurement.

He said: "What exact notion can any man have of such a size as 'a bare sixteenth' or 'a full thirty-second'; and what inconvenient results may ensue from the different notions of different workmen as to the value of these terms. A scale of notation that may have suited the old system of manufacture has been left behind, I am happy to say, as the present age has improved on the past." He advocated a decimal system in which the basic units would be the thousandth and the ten-thousandth of an inch.

It was at this Manchester meeting that Whitworth produced his demonstration that a variation of a ten-thousandth of an inch mattered. He produced a pipe with an internal diameter of 0.5770 inches, and two cylinders with diameters of 0.5770 and 0.5769 inches, and showed that one was tight fit while the other was "so loose as to appear not to fit at all". From the discussion that followed Whitworth's paper, it appears that the members of the IMechE were convinced. James McConnell, ever the enthusiast, proposed a motion "that this meeting pledge themselves to the adoption of the decimal scale, the inch being divided into 1000 parts".

It was never likely to be that simple, of course. But the IMechE returned to the decimal system several times in the course of the next few years, with papers particularly about the benefits in the workshops of the Midland Railway at Derby that the accuracy of the decimal inch had brought.

Whitworth remained heavily committed to the Institution for the rest of his long life, delivering technical papers as late as 1875. He returned to the presidency in 1866, by which time his business was almost entirely in arms manufacture. Two years later he gave the princely sum of £100,000 to set up the Whitworth Scholarship fund, which is still jointly administered by the IMechE and the government. The fund pays towards the university fees of mechanical engineers who have returned to education following a period of practical training such as an apprenticeship. Over the 128 years that the scheme has been running, many of the IMechE's most distinguished members have been Whitworth Scholars.

The IMechE owes Joseph Whitworth a lot. Membership of the Institution, by the mid 1850s when he took over as president, had been stuck at around 200 for half a dozen years. The steady growth in the number of members dates from 1856 – by 1860, it had doubled to 428, doubling again to 855 by 1869. Some of that growth was undoubtedly due to the stature that Whitworth brought to the Institution and the way in which he used the IMechE's meetings as the platform for views and ideas that were taken up nationally and internationally. Whitworth was good at self-publicity, and the Institution benefited from that.

Some of the growth was also due, of course, to the decision to take the Institution's summer meeting away from Birmingham to industrial centres around Britain. The idea of a longer peripatetic summer meeting is credited to James Fenton of the Low Moor Iron Company, who was a member of the council for much of the 1850s. The summer meetings tended from the beginning to be more informal than the set-piece meetings elsewhere, and in the early days large numbers of

Tickets to IMechE events were elaborately designed: this afforded entrance to the 1859 summer meeting at Leeds

In the middle years of the 19th century, the merits of paddle wheels and screw propellers were hotly debated. This tug-of-war in 1849 had a decisive result in favour of the propeller, but the debate raged on for many more years

H.M.S "BASILISK" (COMMON PADDLE) | TOWING STERN TO STERN EACH VESSEL EXERTING HER UTMOST POWER IN OPPOSITE DIRECTIONS. | H.M.S. NIGER (SMITHS SCREW)

papers were delivered, usually shorter and often on a wider range of subjects than the more formal meetings allowed. The 1856 meeting in Glasgow, for instance, heard 19 papers in two days. One of them, by the Sardinian government commissioner Pietro Conti, on iron warships, was the first to be delivered by a foreign engineer.

Summer meetings continued for more than a century, though in the later years the number of papers was drastically reduced and the emphasis was firmly on technical visits, informal contacts and the social aspects.

In the 1850s, Fenton's idea was also that meetings held in different parts of the country would also stimulate local activities, perhaps by showing the mechanical engineers of a region just how many of them there were. In the case of Glasgow, the idea seems to have been borne out immediately: the year after the visit, the Institution of Engineers in Scotland was set up, with close links to the IMechE. Professor William MacQuorn Rankine of Glasgow University, an engineer of vast and long-lasting influence in terms of both education and technology, had been in charge of the organising committee for the IMechE meeting and became the first president of the new Scottish Institution.

Whitworth's presidency of the IMechE – and the Glasgow summer meeting – had a further long-term benefit for the Institution and for the future progress of mechanical engineering in Britain. In his work on precision and measurement, Whitworth was uniting the practical tradition of British mechanical engineering with the science base that was less well-established in Britain. And among the 33 papers at the summer meeting in Glasgow that he presided over was one from perhaps the two most distinguished British engineering scientists of the 19th century: the author was the painfully shy James P Joule, who discovered the mechanical equivalent of heat, and it was delivered by William Thomson, later Lord Kelvin, who also took questions. Their paper, on a surface condenser for steam engines, predates the work they did together which formed the basis of modern refrigeration, though they had already been collaborating on other aspects of engineering science for several years.

The desirability for British mechanical engineering to align itself more closely with engineering science had been identified as one of the conclusions from the Great Exhibition of 1851. The influential Dr Lyon Playfair, writing in November 1851, had issued a stark warning: "As surely as darkness follows the setting sun, so surely will England recede as a manufacturing nation, unless her industrial population becomes much more conversant with science than they now are."

The warning was restated in a different form more than a decade later by William Fairbairn. Following the South Kensington exhibition of 1862, he noted that French and German engineers were ahead of the British in theoretical knowledge. "Not only have they overtaken some of our best efforts in mechanical science, it would now appear the Germans have excelled in the quality of their steel and mechanical workmanship."

The relationship between manufacturing and science was one that would concern the Institution of Mechanical Engineers throughout the 19th century. But there was a further important ingredient in the competitiveness of British mechanical engineering and one that the IMechE had been formed specifically to promote: innovation.

Papers presented to the Institution in its first few years are almost invariably descriptive and deal largely with the industrial development and improvement of existing machines and processes. On occasion, the lack of scientific knowledge led to animated discussion of "advances" which hindsight can tell us were nothing of the sort or meetings managed to hold sensible engineering debate despite scientific error or ignorance.

Famous engineers: from left to right, John Penn, Joseph Whitworth, Robert Napier and William Fairbairn, all of them presidents of the Institution

In 1852, at one of the meetings that Robert Stephenson chaired, locomotive engineer Daniel Clark – later the first locomotive superintendent of the Great North of Scotland Railway – was able to say without contradiction that there did not have to be condensation of the steam during expansion and "in fact there cannot be any except what arises from the abstraction of heat from the steam by external causes". This was a couple of years after Joule's work had been first published, but none of the engineers present, Stephenson included, cited Joule. In fact, the discussion revolved mainly around the merits of superheating, to which the basic scientific error in Clark's paper was perhaps irrelevant.

Probably the first true invention to be unveiled at an Institution meeting came as late as May 1854. The inventor was one of the founding members of the IMechE, John Ramsbottom of the London & North Western Railway. Ramsbottom described the first metallic piston rings – previously hemp had been used for packing. Three grooves a quarter of an inch wide, a quarter of an inch apart and five-sixteenths of an inch deep were made in the piston and fitted with elastic packing rings. The rings were formed from square drawn brass, steel or iron rod that was rolled so as to be 10 per cent bigger than the circumference of the cylinder: they were then placed in the grooves under tension and the whole piston assembly was inserted into the cylinder.

Ramsbottom reported that the new packing had allowed the weight of an 18-inch locomotive piston to be reduced significantly and the fuel consumption of the engines fitted with it was improved by 12 per cent.

The subject of piston rings came up again in 1855, with a demonstration by member David Joy, locomotive superintendent of the Oxford, Worcester and Wolverhampton Railway, of a double coil ring turned from a cylinder of cast iron, which, he said, reduced the wear associated with brass rings. Joy also demonstrated a cramp for inserting the piston, with rings, into the cylinder.

Inventions appeared thereafter fairly regularly in IMechE papers. Also in 1854, Joseph Beattie of the London & South Western Railway described a double-firebox arrangement that he had fitted to six LSWR locomotives. The second auxiliary firebox was mounted for-

Joseph Beattie of the London & South Western Railway described his unusual double firebox locomotives in an IMechE paper

Armstrong, originally trained as a lawyer, later one of the first industrialists to be raised to the peerage

ward of the main firebox and used coal rather than the coke which was standard on steam engines at this stage. William Siemens, like the IMechE founding father Charles Beyer a German emigré, gave the first British demonstration of the gyroscope to the July 1854 meeting: no practical application was foreseen except, perhaps, "as an aid in investigating the action in all parts of mechanism in rapid motion".

Ramsbottom returned to the IMechE with other inventions of his own in both 1856 and 1861: first the steam engine safety valve, then the tender-mounted water scoop that allowed steam locomotives to pick up water from troughs laid between the tracks. Development of the water trough and scoop enabled the Irish Mail to run non-stop from Chester to Holyhead and allowed the locomotive tender size to be reduced so that one extra carriage could be added to the train. During the development, Ramsbottom wanted to know whether the uptake of water from the troughs varied with speed – it did not to any important extent – so he also invented a speedometer.

The late 1850s and the 1860s are important years for mechanical engineering inventiveness, with refinements to locomotives, marine engines, machine tools and other kinds of machinery and the application of precision manufacturing standards. New manufacturing industries, too, were emerging, with guns and other armaments, particularly after the Crimean War and with the involvement of engineers such as William Armstrong in the Institution, forming a link back to the origins of the word "engineer" in military work.

Innovation features strongly in the list of the presidents of the first 30 years. Following Fairbairn and Whitworth, John Penn was a marine engine builder who earned fame from his invention of the lignum vitae bearing for propeller shafts; he returned for a second two-year term as president from 1867.

Armstrong, president like Whitworth for a second single-year term after his initial two-year stint, was the inventor of hydraulic machinery as well as armaments, and a highly influential manufacturer; his

Before making a fortune as an armaments manufacturer, Armstrong had already achieved considerable engineering fame as a manufacturer of hydraulic equipment, such as this hoist

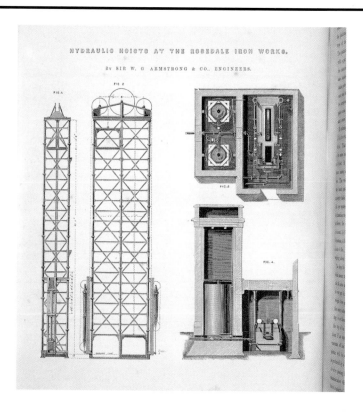

Armstrong-type gun, probably dating from the 1880s

receptiveness to innovation continued late into his long life, for his house, Cragside, at Rothbury in Northumberland was probably the first to be equipped with electric lighting. He was also the first of the great engineers and industrialists to be raised to the peerage.

Innovation in shipbuilding and in the steam engine was also reflected in the IMechE presidency. Robert Napier, the Clydeside shipbuilder who was president for three years in the early 1860s, was credited with the conversion of the Admiralty from wooden hulled vessels to iron hulls in the period just before the IMechE was founded. John Ramsbottom, possibly the most innovative of all the great railway engineers, earned his turn as president in 1870.

There are also gaps in the Institution's lists. While Fairbairn and Whitworth joined in 1847, James Nasmyth, their Manchester contemporary, did not. Isambard Kingdom Brunel, popularly one of the greatest of the mid Victorian engineers, never joined the IMechE, and there is almost no reference to his work in the lists of papers. Though shipbuilding and marine engineering played a large part in the IMechE in the early years, the Leviathan or Great Eastern, Brunel's greatest work or, perhaps, his greatest folly, was not discussed until 1867, when there were papers on the laying of the Atlantic cable by it and on the automatic steering gear which had finally made the great ship manageable at sea. When Robert Stephenson died in 1859, the IMechE went into a sustained ritual of grief – admittedly not as fulsome as that for his father George Stephenson; but when Brunel died in the same year, the event went completely unrecorded.

Equally, though the definition of mechanical engineering appears to have been broad, developments in related areas such as gas engineering and public health engineering were rarely covered. Economic and social effects of the growth of engineering, such as urbanisation, merit no mention until the Institution starts to become a more politically aware organisation from the 1870s onwards, when its presidents begin tentatively making regular speeches on the issues of the day, as Whitworth had done in 1856. Until then, the IMechE is tightly focused on engineering and on the progress of engineering through invention and innovation.

James Nasmyth's steam hammer, invented at the Bridgewater Foundry. Nasmyth was probably the most prominent of the mechanical engineers to turn down membership of the IMechE in its early years. By the time membership was commonplace for engineering manufacturers in the late 1850s, Nasmyth had retired from business

Yet the invention from this period which, perhaps above all others, changed the face of engineering came not from the increasing engineering professionalism that the IMechE represented and was announced at a meeting of an organisation that many felt stood for impractical science rather than practical industry and commerce.

Henry Bessemer, the inventor of mild steel, was a professional inventor with discoveries in glassmaking and textile machinery to his credit. Yet in many ways he was also the inheritor of the old tradition

The Great Eastern, *the biggest ship ever built, was designed by Isambard Kingdom Brunel, who was not an IMechE member, and built by John Scott Russell, who was. It was completed in 1860, after Brunel's death*

The IMechE's interest in the Great Eastern was limited, but in the mid 1860s its role in laying the transatlantic cables, shown here, was the subject of a paper

of the amateur scientist, and the British Association, at whose meeting in 1856 he announced his discovery, was very much the organisation of the scientific establishment.

In fact, the IMechE had allowed Bessemer to escape in 1847. He had attended the second meeting of the new Institution and had, even before that, sent in ideas and models of railway axles. But his notions at this stage received fairly short shrift from Institution members, and he took his ideas elsewhere.

By 1856, the importance of Bessemer's latest work seems to have been absorbed very quickly by the IMechE. Within a few weeks of the announcement, Joseph Whitworth was referring to the invention in his presidential address at the Glasgow summer meeting: it was, he said, "so beautiful and simple as apparently to leave nothing further to be desired".

Bessemer steel process: the IMechE missed out on the announcement, having been less than complimentary to one of Bessemer's earlier inventions

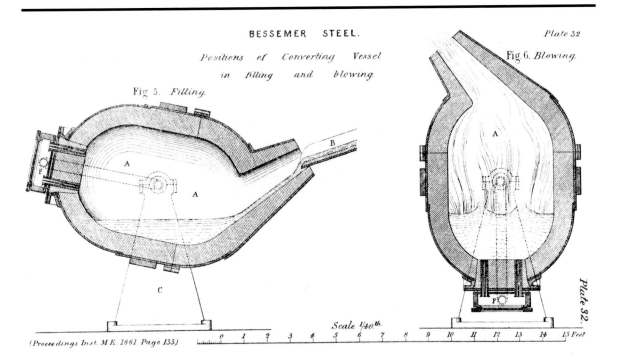

BESSEMER STEEL.

Positions of Converting Vessel in filling and blowing.

Fig 5. *Filling.*

Plate 32

Fig 6. *Blowing.*

Scale 1/40th

(*Proceedings Inst. M.E. 1861 Page 133*)

0 1 2 3 4 5 6 7 8 9 10 11 12 13 14 15 Feet

65

Unfortunately, Whitworth was wrong. The Bessemer converter worked by blowing air through molten pig iron and its very simplicity made it easy to adopt. Within a few days of the British Association meeting, Bessemer had licensed the process to several iron and steel makers, and his own fortune was made. But the process that Bessemer, with limited metallurgical knowledge, had made work ran into problems in the ironworks of experienced founders. The problem turned out to be the sample of iron ore that Bessemer had used and the lining of his converter: Bessemer's ore sample was low in phosphorus, whereas most British iron ores had high phosphorus content, and his lining was a non-siliceous material, while most licensees used an acidic siliceous lining.

The phosphorus problem was not solved for more than 20 years. In the meantime, the Bessemer process was restricted to low-phosphorus ores, available in Britain from Cumberland, Furness and the Forest of Dean. Many overseas steelmakers were similarly circumscribed by the limitations of the Bessemer process, though in some cases they were able to steal a march on their British competitors.

Bessemer finally joined the IMechE in 1861. By this time, the Institution had become very much the forum for the discussion of steelmaking, with William Siemens to the fore.

Siemens had arrived in Britain in 1844 and, after working for Fox Henderson in the Midlands on steam engines and condensers, he set himself up as a consultant in London in the early 1850s, working with his brother. His primary interest was in the regenerative use of waste heat and its application to furnaces. Hot gases were passed through a form of firebrick, which heated up; the gas flow was then turned off and air was passed through the firebrick, becoming very hot. By having two separate chambers containing the firebrick and switching the gas and air supplies between the two, a continuous supply of heat could be maintained.

Siemens presented papers on the regenerative process to the IMechE in 1857 and after the Kensington exhibition of 1862. He saw his process as improving fuel consumption in furnaces of all types, not only in steelmaking but in the glass industry as well and glass manufacturers were much quicker to use it than the steel makers. Another prominent member of the Institution, Edward A Cowper, applied the same principle to hot-blast heating for blastfurnaces, but the first successful steel production in a Siemens open hearth furnace took place not in Britain but in France, where the Martin company had taken out a licence from Siemens in 1863.

With British steel companies still holding back, Siemens decided to go into production on his own account. He opened a small steel-making factory in Birmingham in 1866, and patented the open-hearth process the following year. In 1869, he went into full production of 75 tons a week at the Landore Siemens Steel Company near Swansea. Unlike the Bessemer process, the Siemens method of steel making was relatively slow and therefore controllable. It gained momentum across the 1870s and Siemens, even more than Bessemer, gained fortune and fame from his invention.

Siemens was also elected, in 1872, to join the line of inventors and innovators who had become presidents of the IMechE. He followed John Ramsbottom – who had added to his innovative credentials in

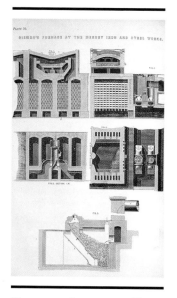

Siemens steel process, as illustrated in a book from the 1870s

Clarke Chapman, later to achieve engineering fame as the promoter of Parsons's steam turbine, exhibited this winch at the Kensington Exhibition of 1862

the 1860s by installing at the Crewe railway works what was probably the first true steel works.

The presidencies of Ramsbottom and Siemens represent a new phase in the maturity of the Institution. They were the first engineers to be elected president on the strength of what they had achieved while they were members, and both had used the Institution to expand and expound their ideas. They were the IMechE's first home-grown presidents.

At the same time, many of the original members were fading away. McConnell left in 1861; Beyer was increasingly ill in the early 1870s, and died in 1876; Slate had died in 1860; Humphrys in 1867.

The Institution's coming of age showed itself in other ways. Membership had increased by 1864 to the point where the Institution's own premises in Newhall Street were no longer big enough for meetings, and the Birmingham events moved to the Midland Institute.

Size was also a factor in the increased clout and a willingness to take on national issues. Across the 1860s, the IMechE took a strong line on the state of the Patent Office, its library and museum. It complained noisily about the difficulty inventors had in finding information on prior patents and the equal problem other people had discov-

Textile machinery such as Platt's mule was still a driving force of engineering technology in the 1862 exhibition. The exhibiting company is Dobson & Barlow of Bolton, whose later head was the exemplary Alfred Dobson (see chapter three)

Crossley Brothers of Manchester was early to spot the potential for the internal combustion engine. It was a Crossley who introduced the topic to the IMechE in 1875

ering information on inventions. The Institution had originally broached the subject back in 1853, but then let things lapse until 1862 when it tried again. Instead of getting better, the Patent Office service got worse, so in 1864 with Napier as president a third broadside was sent, and when this produced no result Napier fired off a brusque letter to the Lord Chancellor. This again took some while to have any effect, and the effect produced was not entirely what the Institution wanted. Instead of the full-time staff demanded, in 1868, there was formal recognition of the Institution's status – or the effectiveness of its complaining – with an invitation to nominate new commissioners of patents for mechanical science.

The patent rows of the 1860s may not have covered the IMechE in glory, but they were evidence of expanding ambition. The ambition would contribute to the decision, finally arrived at in 1877, to move the Institution on to the national stage with a London address.

This was not the only geographical expansion. In 1867, for the first time, the summer meeting went abroad, to the exhibition at Paris, where four of the seven papers were delivered by French engineers and the subjects, including the boring of an artesian well and work on the construction of the Suez Canal, showed a more liberal definition of mechanical engineering than the Institution had encouraged before. The exhibition was also positively the last appearance of James McConnell, who had been so prominent in the Institution's early days. Now a consultant and no longer a member, he acted as one of the judges in the mechanical engineering section.

Had they but known it, the Paris exhibition of 1867 could also have provided the members of the IMechE with a view of one of the engineering devices that would dominate their Institution, their profession and their individual careers over the years to come. It may well have been one of the exhibits that McConnell was called on to judge.

Eight years later, in 1875, Francis Crossley from Manchester described the Otto and Langen Atmospheric Gas Engine, "not only an exceedingly ingenious, but also a well tried and now largely used machine". Crossley's paper is the first description at the IMechE of an internal combustion engine, and his company had already built around 700 of them with outputs of up to 3 hp.

Arthur Paget, a frequent contributor to discussion at the time, remarked that he remembered seeing the engine at Paris in 1867 "and its present appearance is certainly very similar to what it bore then". The president Frederick Bramwell said he was "much struck" at the idea of using a free piston as the device transmitting the force of the explosion: "That seems to be at the root of the merit of this engine," he said. What the 1875 meeting seemed agreed on was that they would be seeing more of this kind of engine in the future.

In that, they were quite right.

3 Moving with the Times

1877 TO 1914

Thomas Hawksley, IMechE president at the time of the move to London and author of one of the most cheerless presidential addresses of them all

Thomas Hawksley's name has long been commemorated at the Institution of Mechanical Engineers in the series of wide-ranging lectures endowed by his son which started in 1913. A generation earlier, Hawksley himself illustrates the fact that the divisions between different branches of engineering were still by no means clear in the second half of the 19th century. He was president of the IMechE in 1876 and 1877, and had been president of the Institution of Civil Engineers five years earlier – had he been born a century later, though, it is a moot point whether he would have been considered as qualified to join either institution.

Hawksley was first and foremost a gas and water supply engineer, and his earliest institutional accolade was to become the first president of what was then the British Association of Gas Managers, the body that would later become the Institution of Gas Engineers. He was a man of forthright opinions and decisiveness. One incident, early in his career, appears to have particularly impressed his contemporaries.

According to his obituary in the proceedings of the Institution of Civil Engineers, Hawksley was the engineer at the gas works at Nottingham at the time of the Chartist riots in the 1840s. The Chartists attacked the gas works, "with the intention of putting the town into darkness". Hawksley organised a barricade which kept

Here and overleaf: The gas industry was well-established by the mid-Victorian period

Victorian gas lamp

them back, then coupled up pipes to the gas supply "and was prepared to play through a nozzle a great tongue of fire on the attacking party in addition to receiving it with shot and hot tar. On those defensive preparations being explained to the rioters, they prudently retired from the works."

The Institution that Hawksley took over in 1876 was about 1,000 strong and had accumulated total assets of more than £11,000. It was wealthy, but in the eyes of outside commentators it was not especially healthy. For them its essential weakness was its location in Birmingham.

The decision to site the IMechE headquarters in Birmingham had been partly accidental: it had been the nearest large town and manufacturing centre to the place where the idea of forming the Institution had been hatched. There was also a practical element to the decision, in that Birmingham, in the early days of the railways, was readily accessible from the north and the south, whereas London was still a long journey – through Birmingham – from Liverpool and Manchester. It is also possible, though nowhere explicitly stated, that the choice of Birmingham was a deliberate decision to distance the new Institution from the older Institution of Civil Engineers, and to assert that a practical provincially-based institution could be every bit as lively and influential as a metropolitan organisation, many of whose members worked in offices rather than factories.

By the early 1870s, what had appeared a reasonable location in 1847 and had been confirmed in 1850 with the acquisition of a permanent headquarters in Birmingham had begun to raise dissenting voices.

The issue grew over quite a long period. In October 1871, the council discussed a proposal that one of the quarterly meetings each year should be held in London. The proposal was still on the table three months later, and decision was again deferred. In May 1872, it was decided to hold a trial spring meeting in London instead of Birmingham to "test the advisability" of London meetings.

Attention at the same council meeting then focused on a new subject: the building committee had found three sites close to the centre of Birmingham which might meet the IMechE's needs for a new headquarters building and meeting hall, the Institution having outgrown its present Newhall Street premises. The secretary, William Prime Marshall, was asked to investigate a plot at the rear of the existing headquarters. He reported back favourably to the July council meeting, which decided "to take steps for securing it at once".

Whether this was done is not clear. At the beginning of 1873, the full Institution deferred decisions on new buildings until the success of the London meeting in the spring was known. The success, when it came, was undeniable: 125 members and 60 guests attended a meeting with two fairly undistinguished papers. This was three times the regular attendance at meetings in Birmingham.

Faced with this kind of incontrovertible evidence of demand for a London presence, the IMechE council did what at this time it seemed to do best: it prevaricated. Of the four general meetings to be held each year, three, it said, should be held in either Birmingham or London – the fourth was the summer meeting usually held in a different centre of mechanical engineering, sometimes abroad.

This indecisive decision worked for 1874, with the spring and autumn meetings held in London. Then, in 1875, it was decided to hold the autumn meeting at neither Birmingham nor London, but at Manchester, where the summer was to be as well. As compensation, the spring meeting in London was to be expanded to two days.

This brought a different problem to the fore. With venues and audiences uncertain, the IMechE was having some difficulty finding enough papers of the right quality to fill the present schedule of meetings; increasing the number or the length of meetings was not an answer. The weekly newspaper *The Engineer*, which at this stage rarely held back when mischievous comment was possible, made unflattering comparisons between the quantity and quality of mechanical engineering papers delivered at the IMechE's quarterly meetings and those which cropped up in the Institution of Civil Engineers' weekly calendar.

The IMechE council tried to rectify matters by announcing two annual prizes, of £100 and £50, would be awarded to the best papers delivered to meetings of the Institution. But this was not the answer.

The answer came, in the end, in something of a rush. The annual general meeting of 1876 installed Thomas Hawksley as president. It also decided to ballot all members – except the council and the overseas members – on where the headquarters should be. Of 834 ballot papers sent out, 574 were returned; 552 voted in favour of London, and just 22 for Birmingham.

It took a year for the ballot to be acted on, but the issue was virtually settled. The 1877 annual meeting debated the motion "that the business and the house of the Institution be removed to London. The debate pointed to some underlying tensions and insecurities within the IMechE. Hawksley's predecessor as president, Frederick Bramwell – later to receive a baronetcy from Queen Victoria – had been one of the authors of the policy of prevarication, but now came out as a supporter of Birmingham. Another former president, however, Sir William Siemens, said he had not been decided before, but was now declaring in favour of London.

Bramwell thought the danger of moving to London would be "injurious competition with the Institution of Civil Engineers". As a "country institution", the IMechE had been allowed to use the Civils' facilities for meetings, but this hospitality would now be withdrawn. The debate had equal numbers of speakers on each side and the show of hands appeared to some observers to be indecisive, but president Hawksley declared the motion "undoubtedly carried". With the overwhelming ballot result behind him, it is difficult to see what else he could have done.

Hawksley also then "bent" the Institution rules by allowing a motion to amend the by-laws and rules to be discussed without the required notice period. The amendments were needed because the original rules stated that the Institution was to be based in Birmingham. Arthur Paget, who proposed that motion, followed it up with a series of other rulebook changes for which he had given notice. They amended the starting time for meetings, allowed members to introduce their friends to meetings, and enabled the council to hold more meetings if the demand was there. The changes, which seem to have been swept through with no discussion, had the effect of lifting some of the self-imposed constraints that had been in place since the Institution was founded.

One small matter was forgotten in all these alterations to the rule book: in November 1877, a further amendment had to be made to delete the words "in Birmingham" from the rule that said the treasurer had to be a banker in Birmingham.

Bramwell was wrong about the Civils' attitude to the IMechE change of location. The spring meeting at the end of May 1877 was held, as it had been in previous years, at the ICE building in Great George Street, Westminster, and the Civils continued to host IMechE meetings regularly for the next 20 years. From the beginning, the ICE had held its meetings weekly: it is difficult to imagine that it would have seen anything to worry about in the arrival in town of a provincial Institution dealing with only an aspect of engineering and composed, in its upper echelons, mainly of ICE members.

Between the IMechE's January and May meetings, there was a lot of work on finding a permanent home in London. A proposal that it should join forces with the Iron and Steel Institute and other smaller or provincial engineering societies in one building was rejected, though it would be revived again with in a couple of years, and then periodically over the next century. At the meeting at the end of May, Hawksley reported that "an excellent set of suitable apartments" had been secured on the ground floor of 10 Victoria Chambers, Victoria Street. The council had appointed a committee to oversee the move and to find a buyer for the Birmingham offices.

The IMechE's first London address was at Victoria Chambers, Victoria Street, seen here in the late 1950s. Within a few years of this photograph, the buildings would be swept away to be replaced by modern office blocks, now housing the Department of Trade and Industry

Two months later, when Hawksley delivered his presidential address to the summer meeting at Bristol, the deed had been done. "You are all aware that the Institution of Mechanical Engineers has exchanged its Provincial for a Metropolitan habitat," he said. "But it may not be known to you that it is already domiciled in Victoria Street, Westminster, where it is in possession of handsome apartments excellently adapted for the transaction of the business of the society, and for the useful accommodation of its members."

The new headquarters, at the Westminster end of Victoria Street on a site now occupied by the glass and concrete building of the Department of Trade and Industry, had "five good rooms" on the ground floor, plus a basement with two further rooms which were to be used to store maps, drawings and publications. The main rooms, Hawksley said, would be used as a members' reading and correspondence room; a library; a secretary's room which would also be the council room; a clerks' or sub-secretaries' office; and a drawing office.

He added: "It is believed that the establishment, thus arranged and properly organised, will become extremely serviceable to Foreigners and Colonists who, having business to transact in this country in relation to matters connected with Mechanical Engineering, have hitherto had no authenticated place of reference for such advice and information as our future Officers will henceforth be readily able to afford." The move, he revealed, had been achieved without asking members for more money and without encroaching on the Institution's reserves, though the new rent would be rather higher.

Hawksley's vision of the IMechE as a thriving business centre may have been about a century premature. But he was right to see the move to London as symbolic of change. The IMechE was about to embark on a new chapter of growth, with a wider range of activities.

The new chapter was to be written without the help of the man who had been secretary for all but the first two years of the Institution's existence. William Prime Marshall decided that, at 62, he would have at most a couple of years in London, and was in any case not keen to move. The council voted him a pension of £400 a year and a lump-sum of £2,000 – and then, not untypically for these disputatious times, had to watch while members took the best part of a year to debate whether this was not being a mite too generous. The payments to Marshall eventually went through in April 1878.

One reason why the IMechE seems to have gone through a period of anxious self-doubt at about the time of the move from Birmingham to London is that mechanical engineering itself was undergoing a few tribulations. Economic historians have had a lot of trouble over the years deciding how and to what degree they can apply 20th century notions about business cycles back into the 19th century. There seems now some consensus that the 1870s and 1880s were a period of recession, if not downright depression, though different industries were affected at different times and to differing degrees.

Hawksley's presidential address in summer 1877 makes it clear that mechanical engineering was going through a difficult time – perhaps the first such recession since railways and factory mechanisation had turned mechanical engineering into a separate discipline with its own profession. Commercial depression, he said, was also matched

New respectability: the IMechE was registered as a company in 1878. Its exact legal status in the 31 years before that is not at all clear

by "marked inventive inactivity".

He went on: "It is true I might have discoursed on the Telephone, the Electric Candle, the Ortheoscope or 'Light Mill' and some other recent developments of physical and chemical phenomena; but however interesting or amusing the descriptions might be made, these and the like inventions and discoveries did not appear to me to be sufficiently connected with our own pursuits to form fitting subjects for discussion on the present occasion."

Not very much later, members might have taken issue with Hawksley over the exclusion of electrical engineering topics as being outside the IMechE's ambit. Indeed, just two years later, the spring meeting of 1879 heard papers on electric lighting and the IMechE was praised for the breadth of its vision by William Henry Preece, one of the pioneers of electrical engineering who was instrumental in forming the Institution of Electrical Engineers out of the old Society of Telegraph Engineers.

At the time, as well, there might have been some reason to cavil at Hawksley's dismissal of mechanical engineering innovation. The same summer meeting heard a paper delivered by William Froude of Torquay on "a new dynamometer for measuring the power delivered to the screws of large ships" which he had built to the order of the Admiralty and which was truly innovative. Froude, a member of the Institution since 1852, had given a paper in 1858 on earlier dynamometers, and had also spoken in 1873 about a machine for making models of ships' hulls out of paraffin wax.

In summer 1877, though, Hawksley had other, more political, concerns, and in introducing two themes that would reverberate across the years, he made one of the first overtly political speeches in the Institution's history.

His first theme was dependence on trade. Britain, he noted, had become through industrialisation a nation that imported a high proportion of its food. In this respect, it was unlike any other country and it was also, therefore, uniquely vulnerable to the disruption of trade routes by unfriendly rivals. This theme – that Britain's dependence on trade and on imports of food made it vulnerable in time of war – would recur many times before the threat became real in the First World War. Hawksley's purpose in raising it at this stage was not just to make this point, though; he also wanted mechanical engineers to help the government change from a policy of building huge warships – enormous and unwieldy floating castles, he called them – to a fleet of light and fast ships that "would succeed in driving every enemy's ship from the face of the sea".

The idea of Britannia ruling the waves was an appropriate one for 1877: it was the year that Queen Victoria was proclaimed Empress of India, and the year from which the British Empire can be truly dated.

Hawksley's other theme that would recur across the years was, again, not a cheerful one. "We are," he said, "at this time diminishing our wealth and the means of supporting our rapidly-increasing population by more than £100,000,000 per annum." This was the extent of the trade deficit caused by the imported food bill, and the deficit was not being reduced, because foreign countries would not take our manufactured goods, which were either too expensive or not well-made. And why, he asked, was this? "Simply because our labour is too dear – too dear in respect of price, too dear in respect of the quantity of work performed and too dear in respect of the obstructions and restrictions which the modern workman thinks fit to place upon his employment and employer."

Parts of the presidential address read as if they were taken straight from the pages of Dickens. Workers, said Hawksley, were under the "transparently false notion" that they were entitled to share their employer's success in business, and had been convinced that increases in wages were permanent additions to their standards of living. The end result of this would be to drive the nation to ruin. "Trades once monopolised by England, the whilome (sic) workshop of the World, have become wholly or partially settled elsewhere, whilst our own working people, still unshaken in their belief in the virtue of strikes, high wages, short hours and workshop restrictions, are only partially employed."

Even for an Institution sunk in unaccustomed gloom, Hawksley's analysis was bleak. The tradition that the presidential address is received without comment was almost broken by Hawksley's predecessor, Frederick Bramwell, who moved the vote of thanks. Bramwell had himself delivered a fairly rigorous presidential address in 1874, exhorting mechanical engineers to take matters more into their own hands. He commented this time that he was sorry "the President had found himself compelled to put before them that which he himself gave as so gloomy a statement", and added that engineers might differ with Hawksley on some points "for if not, they were in a very bad case indeed".

Fortunately, the summer meeting of 1877 at Bristol also included some diversions. One excursion took the IMechE members by free special train and steamer to the Severn Tunnel workings at Portskewett, on the Welsh side of the Severn estuary, returning by steamer – where members were "hospitably entertained at luncheon" – to Portishead docks, which were under construction, and Avonmouth docks. The next day, a second visit from Bristol was arranged, this time by special train to the Swindon works of the Great Western Railway. Train braking was a hot topic of debate at this time,

William Froude described his dynamometer to the 1877 summer meeting at Bristol

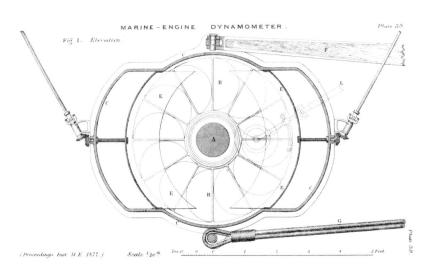

Professor Alexander Kennedy came to prominence in the 1870s as an engineering professor at University College London and as the "reporter" for early IMechE research. He was later IMechE president and knighted for his contributions to power generation, electric railways and engineering education

and the IMechE train was fitted with Sanders's automatic vacuum brake "and its action was illustrated by some quick stoppages at high speed that were made in the course of the trip".

Whether it was the stimulus of the move to London or the apocalyptic vision of their president, there is a strong sense in the IMechE after 1877 that a corner has been turned.

For a start, the Institution decided to put its own affairs on a more regular footing. The exact legal status between 1847 and 1878 is not clear, and company law had been developing over the same period, so there was probably no difficulty. However, the revisions to the rules and by-laws that the move to London involved induced the council to set up a legal entity for the first time. Initially it seems the council hoped to be able work under some form of charter, presumably akin to the Royal Charter that the Institution of Civil Engineers operated under. The Board of Trade turned this down flat but was then sympathetic when the Institution declined to use the word Limited, despite registering as a limited liability company.

The Board of Trade seems in any case to have had a rather difficult time with the IMechE. At one meeting over the company registration issue, the council resolved to seek an interview with the president of the Board of Trade to point out "the great mischief that is continually occurring from interference with the design and construction of Steam Ships and their machinery" and to object to any possible similar interference with the engineering profession.

A further sign of the Institution's reviving impetus was the go-ahead for the programme of mechanical engineering research that had been under discussion for some years. The Civils already had such a programme: the IMechE had decided in principle to sponsor research in 1875, but the proposal lay dormant until Professor Alexander Kennedy, the professor of civil and mechanical engineering at University College, London and an IMechE president in the 1890s, offered the Institution the use of his laboratories.

The annual general meeting in January 1879 was told that the council had agreed to spend £300 "for the purpose of promoting prac-

R H Tweddell was one of the unsung IMechE members who made frequent contributions to discussions in the 1880s and had a solid engineering background with patented machinery that found wide application

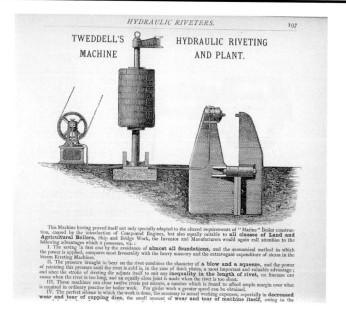

tical research in mechanical subjects". The president, John Robinson, gave three examples of the type of work the Institution might sponsor: "first, the conditions of the hardening, annealing and tempering of steel; secondly, the corrosion of different classes of steel and iron; thirdly, the best form and proportions of riveted joints, both for iron and steel plates."

A research committee was set up to choose the actual topics. Two of Robinson's suggestions made it on to the list of three research projects. The exception was the project on corrosion, and this was replaced by one on friction between solid bodies at high velocities. It was this substitute project which was to have the greatest effect, overturning established scientific theory and leading on, in the second half of the 20th century, to the Institution's pre-eminence in the field of lubrication and friction studies, later, in a moment of 1960s inspiration, renamed tribology. In the council minutes, though, the friction study is very much the third of the three: it would be pursued "should time and money be found to be sufficient".

The other two projects in the first batch of research reflect some of the engineering concerns of the time.

Joints using rivets were being worked on in several parts of industry, including shipbuilding. R H Tweddell, who had spoken at the general meeting that approved the idea of sponsored research, had won a Grand Prix at the Paris exhibition of 1878 for his hydraulic riveting machinery – one of the few British prize-winners at the show. The IMechE's researches led to a report in 1881 that contributed to better understanding of the relations between the net plate section and the bearing surface of the rivets and of the problems of punching holes for rivets.

The research work on steel was to large extent overtaken by events. Mild steel had been gathering momentum since Bessemer's work in the mid 1850s, but despite Bessemer and the later work by Siemens on the open hearth process, the replacement of wrought iron by steel as the basic general engineering material had proceeded slowly. In particular, problems with the high phosphorus content of most British iron ores had limited mild steel applications.

Steelworks in the 19th century relied heavily on manpower: here armour plating for warships is being rolled at the Atlas works in Sheffield

In 1878, at a meeting of the Iron & Steel Institute, a young Welsh amateur chemist named Sidney Gilchrist Thomas, whose professional job was as a clerk at the Thames Police Court in Stepney, announced that he had solved the problem of phosphorus removal: Thomas's idea was to blow the phosphoric iron in a Bessemer converter, thereby oxidising the phosphorus. If the converter lining was composed of a basic material, rather than an acidic material, the phosphoric acid would combine with it to form a slag which could then be removed.

Perhaps unsurprisingly, not a lot of notice seems to have been taken of the claim of this upstart. Thomas however managed to persuade his cousin, Percy Gilchrist, who was a chemist at the Blaenavon Ironworks in south Wales, to try some experiments, and though there were initial problems over the details, the basic principle was quickly shown to be correct. A future IMechE president, the iron and steel maker E Windsor Richards, was quick to seize on the potential of the process and installed large-scale Bessemer converts using Thomas's process on Teesside and in south Wales.

The impact on the British iron and steel industry was huge and very fast. It took much the same time and cost to make five tons of mild steel as it did to make five hundredweight of wrought iron – a twentieth of the amount – and steelmaking could be scaled up and automated much more easily than hand-produced wrought iron. From the start of the 1880s, in Britain, across Europe and in the United States, steelmaking becomes a huge industry, aided by the parallel development of electricity as a power source. The production of iron remained fairly constant, but research work on iron virtually stopped as steel rapidly overtook it in volume.

The speed with which industry's expectations switched from iron to steel – after the long period in the 1860s and 1870s when Bessemer steel was problematic and Siemens' process was being developed – is illustrated by the reaction to the paper on the City of Rome passenger liner given in 1880. The City of Rome was the biggest ship built since the Great Eastern: the decision to build such a prestige vessel using a wrought iron hull was much criticised, and even if that could be justified on ground of cost, IMechE members felt unanimously that steel should have been used for the boilers. One said the use of iron was "most disheartening".

The research work on hardening and tempering of steel was undertaken by William Chandler Roberts – later and better known as William Roberts-Austen. The work followed worries that the colour changes noticed on the surface of steel after hardening and tempering were the result of, respectively, absorption and expulsion of gases. Roberts was able to prove that no gases were involved.

The third of the IMechE's initial research projects, the one on friction, did not get going until 1882. The researcher chosen for the experiments was Beauchamp Tower.

In 1882, Tower was 37, and had been in business as a consulting engineer and designer for four years. Before that, he had worked as assistant to both William Froude, the inventor of the dynamometer, and Lord Rayleigh, who was responsible for developing large areas of engineering science with his work on vibrations and fluid mechanics and who had, in 1879, succeeded the physicist James Clerk Maxwell

Beauchamp Tower: the IMechE's researcher into friction

as Cavendish professor at Cambridge University. Rayleigh was co-opted on to the IMechE's committee on friction at high velocities and may well have suggested Tower as the researcher. Tower had also worked with the past IMechE president Sir William Armstrong on the development of torpedoes.

Tower's own work had produced a spherical engine, but the "great work of his life", according to the obituary of him published in the IMechE Proceedings in 1905, was "the construction of the gyroscopic 'steady platform' for searchlights and guns at sea". This great work, unlike the research work on friction and lubrication, was destined to be a failure. The Royal Navy turned the platform down as too heavy: for the same weight, another gun could be carried on the warship instead.

Tower carried out his experiments over a period of nine years to 1891, though there is evidence that the IMechE stopped funding the work some time in the mid 1880s, and allowed Joseph Tomlinson, chief engineer of the Metropolitan Railway, in whose works at Chapel Street, near Edgware Road station, the tests were set up, to pick up the tab for the later studies. The Institution did, though, publish all four of Tower's reports and 16 years after Tower's final report was published in the Proceedings a new research committee on "the friction of various gears" was set up, with Rayleigh again on the committee.

The breakthrough is recorded in the first report, published in 1883. Tower's testing machine was set up to simulate the conditions commonly found in railway axleboxes, and a variety of lubricants, from lard to mineral grease, was used. He quickly found that "all the common methods of lubrication are so irregular in their action that the friction of a bearing often varies considerably". Consistent results were only achieved when the journal was dipped into or immersed in a bath of oil. This in itself was an important conclusion, because past work in this area had concentrated on friction, rather than on lubrication.

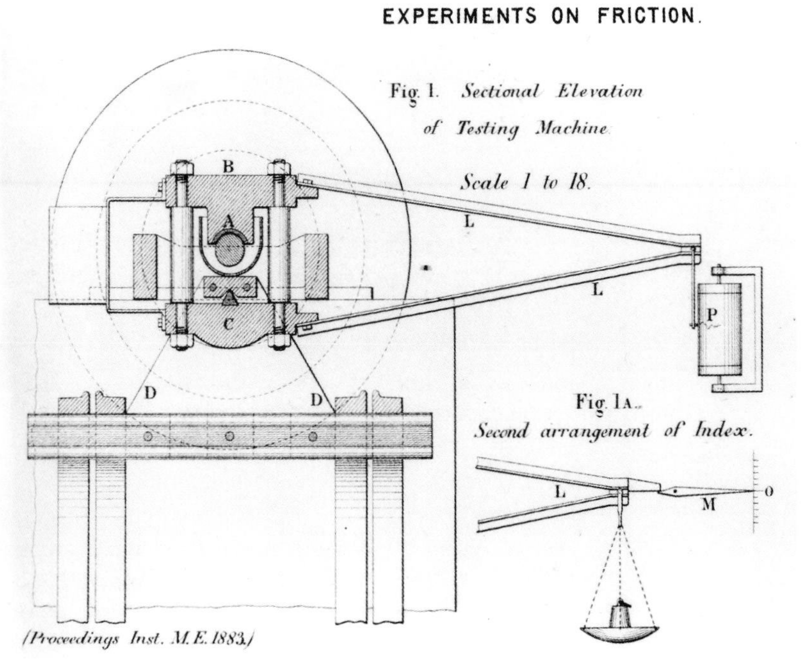

Beauchamp Tower's testing machine for the friction experiments in the early 1880s that overturned previous thinking about lubrication

EXPERIMENTS ON FRICTION.

Fig. 1. *Sectional Elevation of Testing Machine*

Scale 1 to 18.

B

A

C

D   D

L

L

P

Fig. 1A.

Second arrangement of Index.

L   M   O

(Proceedings Inst. M.E. 1883.)

But the real discovery came right at the end of the series of experiments. Tower drilled a half-inch hole in the centre of the bearing. He described what happened next in the report: "On the machine being put together again and started with the oil in the bath, oil was observed to rise in the hole which had been drilled for the lubricator. The oil flowing over the top of the cap made a mess, and an attempt was made to plug up the hole, first with a cork and then with a wooden plug. When the machine was started the plug was slowly forced out by the oil in a way which showed that it was acted on by a considerable pressure."

Tower fitted a pressure gauge into the hole; the gauge was graduated only up to 200 pounds per square inch, and gradually reached that pressure. But the mean load never rose above 100 pounds per square inch.

What Tower had discovered, by accident, was the phenomenon of oil-film pressures; his finding, that the pressure increased towards the middle and diminished towards the edges of the bearing, directly contradicted the prevailing lubrication practice used throughout mechanical engineering and on the railways. Feeding lubricant by hole and groove directly to the centre of the bearing, the standard procedure at the time, "instead of being a means of lubricating the journal, was a most effectual one for collecting and removing all oil from it".

The success of the friction work had a corollary which probably would not have occurred 30 years earlier. Tower's results contradicted established engineering science as well as established practice: in particular, it conflicted with the friction work, a century earlier, of the French scientist Charles Augustin de Coulomb and with the results of more recent trials by the French general Morin, who had been the acknowledged world expert on friction for many years. A generation before Beauchamp Tower, there might have been some chauvinistic glee at the triumph of the British empirical method and practicality over the Continental scientific theory.

But by the 1880s, there was a very different reaction. The Engineer newspaper was among those who called for Tower's work to be followed up, not just by more experimentation, but by engineering scientists. It was no longer enough for British mechanical engineers to know that something worked: they now wanted to know why it worked.

The call for a scientific explanation was not long in being answered. Osborne Reynolds was the professor of engineering at Owens College, Manchester, later Manchester University, the foremost engineering scientist of his day and the man who perhaps more than any other brought together the practical British engineering tradition with the scientific and academic discipline that had been the strength of continental European engineering. In 1886, Reynolds came up with the mathematical proof of the hydrodynamics of oil films and this work in turn led to the development of new types of bearings, new industries and a new branch of engineering in which the Institution of Mechanical Engineers has always played a prominent part, and still does to this day.

A further corollary of Tower's success was that the IMechE's research programme was confirmed, though it continued to operate in a fairly understated way. Committees were set up to investigate marine engines in 1886, steam jackets in 1887 and alloys in 1889, and these were followed by gas engines in 1895, wire ropes in 1913 and cutting tools in 1919.

Most of these, it has to be said, did worthy but not always exciting work. The wire ropes committee had barely defined its remit when the First World War ended research work for the duration, though it finally got going in the 1920s and then continued its work until 1935. The alloys research committee engaged the services of the foremost metallurgist of the day, the renamed William Roberts-Austen, professor of metallurgy at the School of Mines in Kensington, and noted with a degree of justifiable self-satisfaction that the professor's knighthood was the result of his work on alloys of iron, copper and lead for the IMechE. This part of the research was ended only with Roberts-Austen's death in 1902, with his final report published posthumously in 1904, and the research committee then issued sporadic reports on individual projects right up to the mid 1930s.

One benefit of the research reports was that they gave the Institution a continuing supply of subjects for papers. In fact, though, the move to London, the revival of industry from the 1870s recession and the growth of new industries like steel, the expansion of membership both in terms of numbers and of geography and a gradual – and probably unconscious – broadening of the subjects considered suitable for papers meant that the fear expressed in the mid 1870s about insufficient papers to cover even the limited quarterly meetings never materialised. To this day, the IMechE has never been short of material to talk about.

A further factor in this was the increased co-operation with other institutions. Though there is no record of joint meetings as such at this stage, there is evidence that the IMechE was working closely with several institutions of not dissimilar size and status, such as the Institution of Naval Architects, the Iron & Steel Institute and the Society of Telegraph Engineers, which was to become the Institution of Electrical Engineers in the 1880s. The same names crop up in accounts of discussions at all of these bodies, and papers presented seem frequently to be complementary, and rarely clashing or duplicated.

Collaboration might have gone further had a proposal from Sir William Siemens been taken up. In 1879, when the IMechE and some of the others were fairly newly-arrived in London, Siemens offered £10,000 from his own considerable fortune to be put towards a single building to house all the professional institutions covering engineering. The proposal went under the grandiose title of the House of Applied Science.

The IMechE went as far as appointing a special committee chaired by Siemens to investigate the idea and convened a meeting of societies and institutions at Victoria Chambers. Among those invited and attending were the Statistical Society and the Meteorological Society; among those invited but not attending was the Institution of Civil Engineers.

Siemens' idea, it quickly became clear, required the Civils' whole-hearted support as the senior and the richest institution. It also, despite Siemens' own donation and the significant funds of some of the organisations involved, probably needed some support from the government in the form of money, land, or both, and there was some argument that government funding would also be a politically astute recognition of the contribution that engineering had made to Britain's economy.

The Civils continued to make the right sort of noises, without ever committing themselves. As many of the smaller institutions were already using the ICE's facilities for their meetings, it is hard to see what the real benefit to the Civils would have been, and the indications are that the council of the ICE saw any new building as a larger version of their present headquarters, with some common rooms and a lot of space reserved for individual institutions.

In the event, not much happened, and the idea withered away. The possibility of a single building for all the engineering institutions did not resurface until the 1960s.

With a high proportion of the membership living and working outside London, it is not certain how popular the machinations for a grander presence within the capital would have been with the IMechE members. Certainly by 1884, when the Siemens proposal and Siemens himself were both dead, the Institution was facing the first of what would become a regular source of controversy, sometimes complaint.

Professor Robert Smith of the Mason Science College at Birmingham wrote to the president, the iron and steel magnate Lowthian Bell, suggesting that a series of provincial branches of the Institution should be set up in "large centres of mechanical industry". Smith thought the idea "would in my opinion inspire the Institution with new and very vigorous life". He cited Manchester, Liverpool, Glasgow, Leeds, Newcastle and Birmingham as potential centres for branches.

Smith's contention was that the Institution did not punch its weight in terms of official or public recognition. Quarterly meetings with a limited number of papers at each, and discussion that was often cut short, meant there was very little real activity. Members would not travel long distances for so little. Provincial branches would have the effect of increasing the numbers of meetings and of papers dramatically.

Smith's scheme for branches had a rival. Arthur Paget, who had proposed the changes that brought the IMechE from Birmingham to London in 1877, put forward a very Victorian idea for a series of separate regional associations – "cognate societies or institutions with similar objects associated with the institution to which it would pay an annual subscription". Members of these provincial associations would be able use the London facilities such as the reading room or the library for a week if they produced a letter of introduction.

Faced with these competing proposals, the IMechE set up the Provincial Branches Committee, and neatly included both Smith and Paget among its members. The result was stalemate. The committee felt itself unable to make a recommendation on either proposal to the council, so both lapsed. Similar ideas were brought forward several times over the next few years until the question of a regional structure became inescapable – and rather more ill-tempered – in the years before the First World War.

Robert Smith has another claim to institutional fame apart from the desire to change the IMechE's structure. He is an example of how British mechanical engineers were active internationally. It was not just mechanical engineering goods that Britain exported in the second half of the 19th century: there were also the mechanical engineers themselves.

In 1875, Smith was living in Japan, having been selected by Sir Joseph Whitworth to be the first professor of mechanical engineering at the Imperial University in Tokyo – at the age of 22. Smith's appointment was successful, and the department he set up produced the first Japanese-born member of the IMechE. Saku Yokai was a student at the university when he was elected as a graduate member in 1884, with Smith as one of his proposers.

First Japanese members: Tsuneta Shin and Bunji Mano photographed at the 1888 summer meeting in Dublin

Yokai died in 1888 while living in Paris, but had set a precedent. That same year, two other Japanese engineers, Bunji Mano and Tsuneta Shin, visited the IMechE's summer meeting at Dublin and became members. Bunji Mano would later return to Japan to take up Smith's former post as professor of mechanical engineering and to found the Japanese Society of Mechanical Engineers in the image of the IMechE, serving as its first president in 1897.

The Japanese members joined an increasing number of nationalities with the Institution. From Bombay, Nowrosjee Nesserwanjee Wadia was elected as what seems to have been the first Indian member in 1879, though an engineer with an apparently Muslim name appears fleetingly in the lists of members at the end of the 1850s. Wadia was, in any case, joined by Shapurji Sorabji, also from Bombay, in 1887.

The following year, 1888, the Institution president, Sir Edward Carbutt, received a letter from E W M Hughes, the locomotive and carriage superintendent of the Sindh section of the North Western State Railway in Sukkur. Hughes questioned the admission of Indian engineers to the institutions and said that "such persons were apt to gain a footing under false pretences". Hughes added that after 19 years in India he did not know a single native whom he could propose as a proper person to become a member.

If Hughes's target was Wadia, then he had picked the wrong man – and the wrong Institution. Educated in England, Wadia was the son of a civil engineer and had been trained at the Soho Iron Works in Bolton before returning to assist his father at the Royal Mills in Bombay. He returned to England to study paper-making further, and then designed and built several paper mills in India before expanding into textiles.

But Wadia's distinction was not just as an engineer and industrialist. He helped to establish the Victoria Jubilee Technical Institute in Bombay, promoting City and Guilds examinations and becoming chairman of the local Joint Schools Committee. He was a member of the legislative council under Lord Reay, and took a leading role in the Boiler and Factory Acts. He was an associate member of the Institution of Civil Engineers and in 1888, the year of Hughes's complaint, was created a Companion of the Order of the Indian Empire.

To the IMechE council's credit, Hughes's letter was given fairly short shrift. It decided that the election of members should not be dealt with in general, but only on a case-by-case basis as applications were received. On that basis, Wadia and an increasing number of Indian engineers were eminently well-qualified and duly elected.

Membership lists from the 1880s and 1890s show that large numbers of IMechE members were working overseas, not all of them in outposts of the British Empire. There were British mechanical engineers in many European countries and in Latin America, and a few in the United States.

Alfred Dobson was perhaps more widely travelled than most, but there is nothing to suggest his career was extraordinary. Dobson was a partner in a Bolton textile machinery company and his business took him to the United States and Canada, Egypt, India and Japan. His travels in the southern states of the US produced a pamphlet called "A Visit to Cotton Land" and he also wrote texts on various technical aspects of the cotton industry. In 1890 he went to Constantinople to supervise the installation of machinery supplied by his firm at the Yedi Koule Cotton Spinning Mills. At the age of 30 he was awarded the order of Chevalier of the Légion d'Honneur in recognition of his services to French industry.

Travels aside, Dobson appears to have been a paragon of most of the late Victorian virtues of service and example. He was a member of the IMechE council for a total of 10 years and he was mayor of Bolton four times, chairman of the local engineering employers' association and president of the Bolton Chamber of Commerce. He was knighted in Queen Victoria's diamond jubilee honours list in 1897, but died the following year at the age of 50 from pneumonia.

Of course, many members strayed no great distance from home in Britain, but even the least adventurous appear from the earliest days to have expected the Institution to set up and maintain international contacts. And the Institution itself provided some opportunities for travel. The summer meetings were taken overseas several times in the 1880s and 1890s. In 1883, the meeting went to Liège in Belgium, and the president, Percy Westmacott, was summoned to Brussels for an audience with the Belgian king. The 1889 visit to Paris was combined with the exhibition commemorating the centenary of the French Revolution. The IMechE members were entertained at a reception by Gustave Eiffel, whose tower, the tallest structure in the world, was the exhibition centrepiece. They responded by making Eiffel an honorary member.

Increasingly, too, the foreign experiences of both British and overseas engineers formed the subject matter for papers to the Institution. In 1886, for instance, a Russian engineer named Borodin gave a paper on the steam-jacketing and compounding of locomotives in Russia; the following year, there was a paper on the construction of Canadian locomotives; in 1895, a famous mechanical engineering name, Trevithick, was explaining locomotive building in Japan to the Institution.

Within the Proceedings there are occasional extracts from correspondence with French engineers or engineering organisations, and translations are not always provided. Mechanical engineers, like other gentlemen, were expected to know French – though it is not certain how many of them actually did: the first recorded complaint about the British businessman's legendary lack of useful foreign lan-

IMechE members visiting Paris were entertained by Gustav Eiffel, whose tower was under construction as part of the celebration of the centenary of the French Revolution

guages, as distinct from Latin and classical Greek, came in a presidential address in the First World War.

An international dimension of a different sort was the subject of a lecture by T R Crampton, a member of the IMechE council, in 1882. Crampton outlined "a solution of the most prominent engineering problem of the day, namely the proposed submarine tunnel between England and France". This was not the first Channel tunnel proposal of course: Napoleon had had similar ideas more than 70 years earlier and some of the railway builders from the 1850s onwards had devised schemes, in one case making a start on digging.

Crampton's scheme is interesting because he anticipated by more than a century the technique used in the construction of the actual tunnel in the late 1980s. He described a hydraulically driven machine with a rotating head which had sharp-edged discs to slice the chalk as it moved forward. More fancifully, the cut rock would be reduced to a slurry at the cutting face and then pumped away. The resulting tunnel, he said, would be 36 feet in diameter and 20 miles long.

Lectures, unlike conventional papers, were not open to debate, so there was no discussion of Crampton's ideas, and the same restriction applied to the presidential address. In 1882, this may have been a relief. Percy Westmacott, Lord Armstrong's right-hand man in the long years when his Tyneside engineering works was moving from large-scale hydraulic equipment into arms manufacture, used his address to come up with an outrageous scheme for the dockside unloading of ships. He had seen how huge cranes lifted coal barges out of the water and decanted their contents on to wharves, and reasoned that ocean-going steamships might be dealt with in the same way. They could be "lifted bodily out of the water, and then scuttled of their contents right and left into the warehouses, at a rate and with a saving of labour that would far exceed anything hitherto attempted". A secondary bonus would be that shipowners could inspect their hulls more easily.

The idea was, of course, an exaggeration, though, to be fair to Westmacott, shiplifts and major canal works were an intermittent theme in the period right up to the First World War. The Suez canal, completed in 1869, had been acquired by Disraeli for the British in 1875; the French tried, unsuccessfully, to build a canal across the Panama isthmus in the 1880s, and the Americans later succeeded; in Britain, the Manchester Ship Canal was completed in the 1890s.

More consistent themes in the papers delivered to the IMechE in the 1880s and the 1890s were steel, with bigger and more efficient plants being built continuously, especially in the United States, and the wonder of the age, electricity.

The Society of Telegraph Engineers had transformed itself into the Institution of Electrical Engineers in 1881, but the compartmentalisation of engineering into component disciplines was far from clear, and in the area of power generation at least it never would be, with mechanical and electrical engineering inextricably intertwined. The IMechE was not averse to straying into territory that was more definitely the preserve of the electrical engineer on occasion. There were papers, for instance, on the electrical equipment of a Scottish lighthouse and the Bolton traveller Alfred Dobson presented a paper in 1893 on how electric lights made hands work at his factories.

Disaster and triumph. The Tay Bridge disaster in December 1879 was the result of using materials and a structure unsuited to the loads placed on it; just 10 years later, the Forth Bridge showed the lessons had been learned

Sir Charles Parsons, inventor of the practical steam turbine

The crossover between the mechanical and the electrical engineer was represented by the steam turbine and the man who turned an idea that had originated with the ancient Greeks into a practical power source, Charles Parsons.

Parsons is an example of how the long British tradition of amateur science was, by the end of the Victorian period, fusing with the newer practical engineering skills. The son of the Earl of Rosse, president of the Royal Society and owner, at Birr Castle in Ireland, of the largest telescope in the world, Parsons had a traditionally privileged upper class upbringing, including studying mathematics and physics at Cambridge. After leaving university, though, he was sent by his father as a student apprentice to Armstrong's factory at Elswick on the banks of the Tyne to gain practical engineering experience. From Armstrong's, he moved to another Tyneside company, Clarke Chapman, which was manufacturing marine engines but which had also taken an interest in electric lighting and the power generation equipment it required.

Parsons had begun experimenting with rotary steam engines while at Cambridge. At Clarke Chapman, he became convinced that a rotary engine was a better bet for driving a dynamo than the marine steam engines the company was famous for. In 1884, he took out a patent for a steam turbine.

Parsons's first turbine generator, patented in 1884

Unlike the Swedish engineer de Laval who was also working on a single-stage turbine design at this time, Parsons's turbine was a multi-stage device with alternate fixed and rotating blades through which the steam was gradually expanded. The original turbine produced an output of just 10 hp at 18,000 revolutions a minute and by 1888, when Parsons presented the results of his work to the summer meeting of the IMechE in Dublin, the machines had been scaled up by a factor of about 40, with the maximum output so far 32 kW. Clarke Chapman had made about 360 small turbo-generators for power generation at this stage, including some for shipboard use.

Despite the limited output, IMechE members seem to have grasped the significance and potential of Parsons's invention immediately and Parsons himself was fairly free with his information. Experienced engine builders noted that the turbine's efficiency was already impressive, and Parsons said that if a condenser was added he

Portable turbine generator from 1886: this is on its way to generate power to illuminate electric lights at the lake at Sir Joseph Swan's house at Gateshead for skating

Parsons split with Clarke Chapman over the use of larger turbine units in power generation: this is Forth Banks power station at Newcastle in 1890

By the turn of the century, Parsons's own turbine manufacturing plant was a sizable operation

expected the turbine to outperform any reciprocating engine in terms of fuel economy.

In fact, there was so much discussion on the paper that the session had to be resumed in London in the autumn. Among the ideas that emerged was one from past-president Jeremiah Head, who suggested that the turbine's exceptional power to weight ratio might make it suitable for "aerial navigation" – and went on to speculate that it might be possible to drive a turbine by the hot gases from the combustion of petroleum, thus doing away with the boiler entirely. No-one seemed to recall that the impossibility of rotary engines had

The ship that shocked the Royal Navy: the Turbinia *was the sensation of the 1897 review of the fleet*

been the subject of George Stephenson's only paper to the Institution of which he was the first president.

In contrast to the IMechE members and Parsons himself, Clarke Chapman's ambitions for the steam turbine were fairly limited, and Parsons left the company in 1889 to set up his own company. Because of a patent dispute, however, he was unable to make axial flow machines for some years. By 1900, however, he was manufacturing turbines with outputs in excess of 1 MW in his factory for electricity generation.

By this stage, of course, Parsons and his turbine had an additional claim to fame. In the 1890s, he started work on adapting his turbine for marine propulsion, and in 1897, at Queen Victoria's diamond jubilee review of the fleet in the Solent, his turbine-powered launch the *Turbinia* created a sensation by racing among the Royal Navy's ships at an astonishing 35 knots. The *Turbinia* had three turbines each driving a separate shaft with three propellers on each shaft; the power output was 2,000 hp.

The adaptation of the turbine for ship propulsion contributed to the revolutionising of naval warfare in the first decade of the 20th century. The turbine allowed naval designers to use speed as a weapon for the first time and it could be allied to developments in ship size and in the number and calibre of heavy artillery. The first battleship to combine all of these features was the *Dreadnought*, launched in February 1906. At a stroke, said one commentator, all other warships in the world were rendered obsolete.

The IMechE had something of a ringside seat for the fierce debates within the Admiralty that led to the *Dreadnought* and the rise of awesome naval power. Sir William White, the last of the great Victorian naval designers and the builder of the King Edward VII, the Royal Navy's big battleship delivered just a couple of years before the

Dreadnought, was president of the Institution at the turn of the century and an influential figure within the IMechE for more than 10 years after that.

The turbine was one of several mechanical engineering technologies that would take the IMechE forward into the new century. From the discussions of the 1880s and 1890s, there seemed to be no shortage of other promising candidates. Probably at no stage in the Institution's history has the definition of mechanical engineering been so wide.

In 1882, for example, there were papers on the Laval centrifugal cream separator and on the first "breathing machine" to allow men to work in poisonous gases. The French Decauville instant light railway, consisting of preformed track and sleeper sections that could be laid on even ground by unskilled labourers, was described by its inventor in 1884; he forecast that his railway would provide the first service between England and France. Bicycles and tricycles were of constant interest: one paper presented to the IMechE in 1885 had 122 illustrations of different types, with the penny-farthing apparently being challenged by newer designs in the two-wheeler market.

By the early 1890s, power station engineering was taking up a lot of IMechE time, aided by the election of Professor Alexander Kennedy.

Victorian pleasures: the bicycle in all its forms was the subject of a comprehensive paper at the IMechE in 1885

Fig. 80.   *Tandem Tricycle.*

(Proceedings Inst. M.E. 1885.)

Kennedy was the leading authority on power stations, the head of Westminster's electricity department, the consultant engineer for the electrification of the Metropolitan Railway, and, for good measure, emeritus professor of engineering at University College London. His view, which he lived long enough to see entirely overturned, was that electric lighting provided the only real justification for power station construction: industry might be able to use electricity in some areas if factories were sited conveniently, but would struggle to justify the installation of generating capacity for its own sake.

The healthy state of debate within the IMechE was reflected in healthy membership and finances. In the 15 years after the move from Birmingham to London, membership doubled to more than 2,000; the Institution's income went up in step with the number of members, but because the activities remained much the same from year to year, the profits rose substantially. By the mid 1890s, the Institution was seriously wealthy, with investments totalling more than £27,000.

In these circumstances, a rented set of five rooms and a basement in Victoria Street, plus the continued use of the Institution of Civil Engineers for meetings in London, was unnecessarily modest. In 1894, the council set up a buildings committee to look for a suitable site for a proper headquarters building. The summer meeting of 1895 heard that a lease had been granted by the Church Commissioners for a site at Storey's Gate, Westminster, facing on to Birdcage Walk. With St James's Park across the road, the Treasury building opposite and the Institution of Civil Engineers just along the road, it was one of the grandest addresses in London.

The following year construction work started, to a design by Basil Slade. The accommodation included a large lecture hall, a library to hold the 10,000 books and documents that the Institution had accumulated, a reading and smoking room, a council chamber and a tea room, plus offices for the staff. Instead of raiding the corporate coffers to pay for the work, the Institution invited members to subscribe to £100 debentures, offering to pay 4 per cent interest in return. The aim was to raise £25,000 by this means; the scheme was oversub-

A home of its own: Slade's designs for the new headquarters of the institution at Storey's Gate

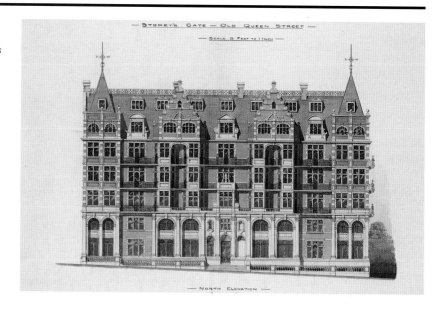

scribed, with 75 applications for £52,700 worth of debentures. The Institution was wealthy, and so were some of its members.

The year 1897 was the diamond jubilee of Queen Victoria; it was also the golden jubilee of the IMechE. The two events were joined together in a loyal address sent to the Queen under the seal of the Institution. It said: "We are deeply sensible of the wise and beneficent sway which your Majesty has throughout exercised over your Empire, and of the great advances consequently realised in Art and Science, particularly in Mechanical Engineering, to which this Institution, having been established ten years after your Majesty's accession to the Throne, has now been devoted for half a century. We earnestly pray that your Majesty may still be spared to reign over a faithful and affectionate people."

With the new headquarters still under construction – and running into trouble because of the discovery that the site at Storey's Gate stood on a small island surrounded by deep water – the main jubilee celebrations were held at the Institution's birthplace, Birmingham. The president, Edward Windsor Richards, the steelmaker who had done much to encourage the Gilchrist-Thomas process, solemnly read out the loyal address and the reply from the Home Secretary, and then devoted his address to a review of the preceding 50 years.

Of the founding members who had met at the Queen's Hotel in Birmingham on January 27, 1847 only two were still alive – John Ramsbottom had died two months before the 1897 summer meeting, leaving only Peter Rothwell Jackson, the inventor of the wheel moulding machine who had described his invention in a paper as long ago as 1855, and Richard Williams of Wednesbury. Jackson, who died in 1899, was not at the meeting; so Williams was prevailed upon to respond to the president's address.

Williams was well into his 70s, and began nervously. He had been a member, he said, for 50 years and had attended most of the early meetings, but he had never before had to make a speech. He paid tribute to William Prime Marshall, the secretary from 1848 to 1877 who was now an invalid, and reminisced about George and Robert Stephenson: "In my memory they are both as vividly present to me now as they were when they addressed the Institution in its early days." George Stephenson in particular had had "peculiar methods and peculiar ideas, which he expressed in peculiar language, but they were eminently engineering ideas, and specially calculated to give vitality to the Institution".

Williams died in 1909, outliving all the other founding members by 10 years. There is no record that he made any other speeches.

The construction delays meant that the new headquarters in Birdcage Walk was not ready for occupation until 1899. The formal opening in May was celebrated with two evening receptions, attended by more than 1,200 members and guests, each of whom was presented with a brochure describing the new building and giving a brief history of the Institution. The London Concert Orchestra played, the Meister Glee Singers sang, and members were treated to the first Institution demonstration of kinematograph films. The Daily Telegraph's correspondent reported that the building had the atmosphere of a club rather than an Institution building. It is thought the remark was intended as a compliment.

The move into the new headquarters allowed an expansion of activities and a thorough review of the Institution's procedures. A committee structure was set up to handle administrative tasks such as scrutinising ballot papers for elections and recommending technical papers for reading at meetings. The number of meetings almost instantly increased to "monthly": in practice, this meant eight times a year, instead of the four that had been in place since 1847.

Probably the most important of the new committees was the creation of a separate graduates' committee, a direct initiative by the president Sir William White. Graduate meetings were to be held fortnightly and were intended to help develop younger engineers as members; as well as lectures and papers, the graduate section from the beginning included regular visits to factories and other engineering installations, as the senior members tended to do on their summer meetings. In the first year, visits were arranged to the London Hydraulic Power Company, whose power network included the mechanisms for raising and lowering Tower Bridge, the Woolwich Arsenal, the Midland Railway works and the Central London Electric Railway. London graduate members inevitably dominated, but the first committee also included members from Birmingham, Manchester and Cardiff.

White himself chaired the first meeting, but the committee thereafter exercised considerable independence – and enterprise. The second paper ever read to the graduates' section in 1899 was on "Motor haulage on common roads" by Alfred Marsden from Leyland in Lancashire. This was essentially a review of light steam cars, but discussion ranged over petrol vehicles and wheel types. Marsden was back little over a year later with a second talk on motor-car transmission mechanisms, and the audience for this second paper included the engineer and racing driver S F Edge and Charles Stewart Rolls, the motoring pioneer and balloonist who would, in 1910, be Britain's first aviation casualty.

Charles Stewart Rolls complained to the IMechE in 1900 about the innate prejudice of the English against the motor car

British industry was far from convinced that the internal combustion engine would triumph over steam: this is a Stanley steam car of around 1899, with early pneumatic tyres

Herbert Austin was responsible for Wolseley's move into cars: his wife is seen here in an early Wolseley

In the discussion following the talk, Rolls spoke about the distress of the British motor trade "suffering from the effects of the innate prejudice of Englishmen". Exactly what he was referring to is not clear: there is no doubt, though, that many of the early motoring pioneers in Britain felt frustrated at the legislative and infrastructure restrictions on motoring, at the slowness of manufacturing industry in Britain to seize the opportunities presented by the new technology of the motor car, and at the caution shown by many in Britain, engineers among them, in recognising that the petrol engine, rather than the steam engine, would be the driving force for a whole new branch of engineering.

In this instance at least, the impatience of Rolls and the graduate members of the IMechE may have had justification. The juniors beat the seniors by a full year in discussing the new automobile industry – the first paper on motor cars delivered to the Institution proper did not appear until 1900, when the future president Dr H S Hele-Shaw read a paper on road locomotion.

But frustration can be seen over a longer term as well. The 20 years from 1890 to 1910 saw, in many ways, as dramatic changes in engineering technology and industry as the 20 years from 1830 to 1850, which had seen the initial development of mechanical engineering spearheaded by the railways. But where Britain and British engineers had led the way in the earlier period, there was now an uncomfortable and incontrovertible feeling that other countries were setting the pace. Several factors were blamed: an innate and very British conservatism, elderly equipment, the diversions of an imperial role, industrial relations difficulties. If no-one as yet blamed the engineers, it was only a matter of time.

British conservatism on the development of the automobile had been addressed as early as 1895 by the IMechE. At the autumn meeting that year at the Royal United Services Institute in Whitehall, the president Professor Alexander Kennedy "called attention to the memorial which was being prepared for presentation to the Right Honourable Henry Chaplin MP, President of the Local Government Board, urging the repeal of existing statutes so far as they operated to prevent the use of light vehicles propelled by steam or other motive power and not employed in traction".

By the time of the second IMechE paper on cars, presented by M Holroyd Smith of London at the summer meeting of 1901, held at Barrow-in-Furness, a degree of self-criticism was creeping in. "Perhaps there is no mechanical subject at the present time occupying the serious attention of a larger number of engineers," Holroyd Smith said. But, he added: "Many motor vehicles produced by English makers consist of numerous parts collected mostly from France and assembled together here. From a business standpoint this may have been a prudent course to adopt, but it is hardly creditable to the English engineering profession."

The criticism about Britain's slowness in developing its own motor manufacturing industry died down as production picked up across the first decade of the new century. But it never entirely went away. Britain was an easier market for American transplants such as Ford to enter not just because of language and historical affinity: it also had weaker home players.

The other great transport development of the years around the turn of the century, manned flight, seemed to cause less anguish to the British, partly because it was, in the period up to the First World War, less obviously a progenitor of a new industry, and partly because British engineers had made a material contribution to the science that underpinned the new technology.

Among the engineers who had worked on aspects of flight in the

The trademark of summer meetings in the years up to the First World War was the group photograph: this is part of the 1901 meeting at Barrow in Furness, taking the air at Furness Abbey. The caption to the picture identifies the back row as, from left to right, Mr Pettigrew, Mr Sire, Mr Stileman, Col Strongitharm and Mr Donkin. The front row is, from left to right, Mr Ivatt, Sir Hiram Maxim, Mr Maw, Mr Aslett, Mr Chapman and Mr Matthews. H A Ivatt was the chief mechanical engineer of the Great Northern Railway; Sir Hiram Maxim was the armaments inventor and entrepeneur; William Henry Maw was president in 1901, editor of Engineering *and a constant presence at the IMechE until his death in 1924*

1890s was Charles Parsons. The turbine pioneer was researching the effects of jacketing steam engine cylinders in 1893 – in parallel with the IMechE's own research on the same subject – when he tried to make model engines sustain their own weight in the air by propelling an airscrew. He had some success with toys of wingspans up to 11 feet, but concluded that "the problem of aerial flight can be attacked much more favourably by means of cigar-shaped balloons propelled by gas or oil engines".

The American-born armaments entrepreneur Sir Hiram Maxim, who had presented a paper to the IMechE on his automatic machine gun in 1885, was another to try his hand at powered flight in the 1890s. He built a full-size biplane powered by steam engines and to be launched from rails. In tests in 1894, it briefly lifted from the rails, but it was too complex and too heavy to really work.

Another pioneer was Frederick Lanchester, later known better for his cars. Lanchester flew model gliders from 1892, and devised what he called the "vortex theory of lift", which he explained in a paper to the Birmingham Natural History Society in 1894. He intended that a more detailed version of his researches should be published by the

F W Lanchester's flying machine studies anticipated the first flight by the Wright Brothers in 1903, and continued for many years afterwards

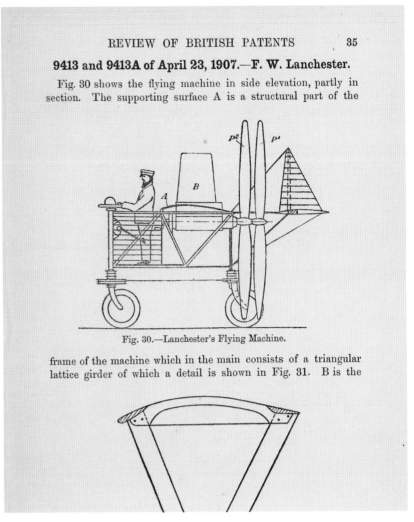

REVIEW OF BRITISH PATENTS   35

9413 and 9413A of April 23, 1907.—F. W. Lanchester.

Fig. 30 shows the flying machine in side elevation, partly in section. The supporting surface A is a structural part of the

Fig. 30.—Lanchester's Flying Machine.

frame of the machine which in the main consists of a triangular lattice girder of which a detail is shown in Fig. 31. B is the

Royal Society or the Physical Society, but they were not interested. It was only after the first flight by the Wright brothers in the United States in 1903 that there was enough interest to justify publication. Lanchester's thinking eventually extended to the two-volume book *Aerial Flight*, published in 1907/08, which is the first definitive account of aeronautical science and engineering. Almost 40 years later, in 1945, at the end of a war in which aerial power had been decisive, the IMechE awarded Lanchester the James Watt International Medal, citing this book, rather than his automobile exploits, as his crowning achievement.

In truth, though, the IMechE was scarcely in the forefront of aviation technology in the early days of manned flight. The first paper on aerial navigation was delivered by Dr H S Hele-Shaw at the president's conversazione in 1907, a remarkable double following his earlier paper on automobile engineering in 1900. The aviation paper was accompanied by a cinema film of flight attempts. And if it took nearly 40 years to honour Lanchester, the surviving Wright brother, Orville, fared no better at its hands: in 1942, just 39 years after the first flight at Kitty Hawk, he was elected an honorary member.

The transport innovations of the 20 years to 1910 and the fact that British industry and British engineers were conspicuously no longer leading the way caused something of a crisis of confidence in the years leading up to the First World War. The popular view of the Edwardian age is heavy influenced by our knowledge of what came next. It is an E M Forster view of a serene golden age of elegance, of endless sunshine, cricket matches and codes of honour, confidently played out by a generation that we know to be doomed.

In practice, for many in industry Edwardian Britain was more about grim industrial relations problems, increasing social legislation placing an uncertain burden on employment, the growth of protectionism and the rapid industrialisation of competitor countries – plus the accelerated rate of technological change. For industry it was a difficult period. And the IMechE on occasion reflected that difficulty.

In 1904, for example, the Institution's summer meeting was held, for the first time, in the United States. The delegation's American hosts laid on the hospitality to an inordinate degree: in one account, a total of 191 engineering works and offices in the Chicago area were said to have been opened to the British engineers. What the engineers saw impressed them.

Summer meetings, particularly at that distance, were attended by only very few of the IMechE's members. Considerably more will have seen the paper, reproduced in the Proceedings for January 1905, by A J Gimson of Leicester on American workshops. Gimson's view was that American workshop practice was more single-minded that British, with the result that productivity was greater and workers' skills were used more effectively. The whole of the US engineering factory's efforts, he said, were focused on economy and accuracy.

Gimson's analysis patently annoyed some IMechE's stalwarts. Charles Wicksteed, from the Kettering company whose equipment is found in children's playgrounds across Britain, said that he had "right or wrong told Americans that English engineering shops were better equipped than their shops were". He said: "I like to stand up for the Old Country."

But IMechE council member Mark Robinson conceded that US production was better organised. And another member, W Stanley Bott, suggested that American cost control had much to teach the British.

Gimson's paper and the responses to it are not untypical of a new strand emerging in the debates at the IMechE in the first decade of the 20th century. The strand is self-critical without being over-critical. It proceeds from a recognition that Britain's lead in many sectors of industry has been eroded, that technology is international and that competitiveness – needed now as never before – requires engineers to exercise organisational and managerial skills, as well as technical abilities.

As early as the summer meeting of 1901, there had been papers on workshop organisation and the motivation of employees through payment schemes. William Weir of the Weir pumps group in Glasgow explained, with a colleague, schemes such as the Workmen's Suggestion Scheme and the Friction Club – for foremen to get together to discuss problems – and how these contributed to factory efficiency. The growing implication across the decade was that technology on its own was not enough.

In addition to these pressures, the Institution was also facing new competition in its representational role. New institutions for water-

In 1904, IMechE members ventured on a visit to the United States and Canada, a world of big buildings and vast industrial sites. Their impressions of American industrial efficiency were a potent factor in the drive for change within the Institution

works engineers, heating and ventilating engineers, automobile engineers, structural engineers and locomotive engineers were created between 1896 and 1911. The IMechE too was growing fast in these years, but there was a growing need for the Institution to define more closely what it meant by "mechanical engineer".

The presidential address of Edward Ellington in March 1911 represents a culmination of the first stage of a process of adjusting to these new realities. Ellington, a hydraulic power specialist, eschewed the

The first examination paper for graduate membership in 1913 required candidates to demonstrate a broad range of knowledge

The Institution of Mechanical Engineers.

GRADUATESHIP EXAMINATION.

GENERAL KNOWLEDGE.
TUESDAY, 7TH OCTOBER 1913.   (10.0 A.M.–1.0 P.M.)

Essay obligatory ; also ONE *question may be answered in each of the sections* A, B *and* C.

The maximum number of marks obtainable for each question is shown in brackets.

1. Write an essay on *one* of the following subjects :—
   (*a*) Historical novels.
   (*b*) Recent scientific discoveries.
   (*c*) Engineering in connection with aviation.   [55]

A.

2. Put the following writers' names in order of date ; mention some work by each of them, and write a short account of any *one* of the works you specify : Bacon, Chaucer, Macaulay, Spenser, Tennyson.   [15]

3. Write a character sketch of any important figure in a Shakespearean play.   [15]

B.

4. Give some account of British possessions *either* in Asia, *or* in Africa.   [15]

B 2

usual presidential practice of speaking about his sector of industry. He did make a few domestic announcements, including the news that the lease of the Storey's Gate Tavern, which occupied the plot next door to the IMechE headquarters, had unexpectedly come vacant and had been acquired by the Institution for an extension – needed because membership had advanced from 2,700 to 5,700 in the dozen years since the new building had been built.

But the bulk of Ellington's address is a carefully worded exposition of the progress of mechanical engineering from its earliest days, interwoven with a history of the Institution, out of which came a clear message. Engineering had progressed, and the Institution that served it had to progress too, providing industry with engineers who had proven qualifications and qualities which would include technical and scientific knowledge as well as managerial and economic skills.

"The Institution is intended for mechanical engineers," he said. "The test for admission should be high, both from the point of view of general and technical education and of practical work. It should always be kept in mind that the members of this Institution ought to have not only a good theoretical and practical knowledge of mechanical engineering, but should be prepared for the control of engineering industries and the workmen employed in them."

The following February, the draft scheme for the first IMechE entrance examinations was published in the proceedings. Candidates for associate membership would be expected to sit a general knowledge paper, one three-hour paper each on applied mathematics and on physics and chemistry, and two three-hours papers selected from a list that included materials, steam engines, internal combustion engines, hydraulics, "electrotechnics", the theory of machines, machine design and metallurgy. Possession of a university degree or a City and Guilds diploma would be accepted as an exemption. The first examinations were held in 1913.

Other changes followed. A petition of members at the end of 1910 called for the Institution to set up a more regular form of communication with its members, in the form of a journal that might supersede the Proceedings. A ballot supported this, and though the 1913 president, Sir Frederick Donaldson, announced that the support was insufficient to justify the cost of publication, the first journal duly appeared in 1914.

The question of regional branches also reappeared. Future president Daniel Adamson, in a speech to the 1912 annual meeting that was outstandingly and wide-rangingly complaining, grumbled about the lack of research programmes and then read out a letter purporting to have come from "a friend" which complained that members outside London were ignored. "Increases in membership are not sufficient rejoinder to the criticisms," he said. "The Institution is putting on fat rather than muscle."

IMechE summer meetings at home and abroad in the years around the First World War were commemorated by badges

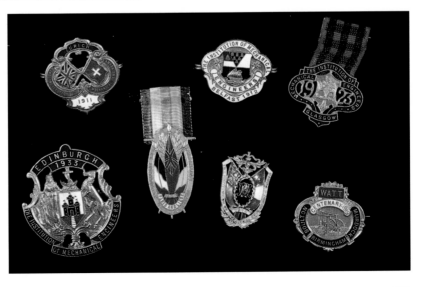

The regional question was shelved again, but there was change in the air. The Thomas Hawksley lectures, set up from 1913, were deliberately repeated at regional centres, and unofficial discussion groups also met in the regions to debate papers delivered in London.

While all this change was afoot, the day-to-day business of the Institution continued. Highlights included a paper by Rudolf Diesel on his engine, in which he claimed that the diesel engine was especially important for Britain because it would help to conserve dwindling coal stocks. And in 1914, the summer meeting was once again fixed for Paris. The political storm clouds had been gathering over Europe for a while when the IMechE party set out in early July for the French capital. Following two days of technical papers and receptions, the meeting broke up into groups for visits to French factories. By the middle of the month, the mechanical engineers were back in Britain. Three weeks later, Europe was at war.

4

Battles lost and Won

1914 TO 1939

The First World War is perhaps also the first watershed in world history. Before it, no other event, or series of events, had ever changed the whole world.

By comparison with 1914 to 1918, many of the most famous dates in history had repercussions that were limited by geography or which unfolded over a considerable period. The year of the Norman conquest and the suppression of Anglo-Saxon England, 1066, is a watershed in British history and in parts of Western Europe, but not necessarily elsewhere: historians of Spain or Byzantium would reckon the year before, or the year after, was politically more momentous. Columbus may have sailed the ocean blue in 1492, but the development of the American continent took a few centuries after that.

The 1914-1918 war, though, affected people's lives around the globe in a way that only events with religious significance can claim to rival. More than that, the war affected each and every aspect of the lives of individuals and of the societies they had created. It caused economic shock waves that are felt to this day; it upset established social orders, with consequences that could never be reversed; it changed the political map of the world. In Britain, it caused, or led indirectly, to events as diverse as votes for women, the partition of Ireland and restricted licensing hours for alcohol sales.

It was also, by previous standards, carnage on an unprecedented scale. Never before had casualties in a war been numbered in millions; never before, in wars that Britain had been involved in, had so little gain been made at so great a cost. Britain had been detached from wars in Europe against other European countries for almost a century, with the exception of the Crimea. Small-scale colonial wars, often in distant parts of the globe against ill-equipped locals, were no preparation for the economic scale nor the geographical proximity of the war in Europe. And Britain had stood aloof as other nations raised the stakes in savagery in wars such as the Franco-Prussian or in far-off areas such as Armenia or the Balkans, where civilian massacres extended the definition of combatants.

So being at war was a shock, the nature of the war was a second shock, and the shocks kept on coming, for four long years.

In the early days of the war, engineers and their institutions appear to have been swept along with the national mood of enthusiasm and optimism. They had particular reasons for being involved. From the outset, the war was seen as a trial of technologies and military hardware that had not been available in previous conflicts or whose use had been limited by geographical distance. Heavy armaments, shells and mines represented relatively untried ideas; motor transport had

The scale of war's victims was enormous in every sense; this is a propeller from the 31,000 tonne Cunard liner Lusitania, *sunk by a German submarine in 1915 with the loss of 1198 lives*

not been available to move troops around quickly and in large numbers before; telephones and telegraphy provided fast communications between the frontline troops and the generals further back, and between the generals on mainland Europe and their political masters in London; at sea, traditional surface warfare had now been joined by the potential for submarine subterfuge.

As the war went on, further innovations were put into practice. The first tanks, with strange-looking all-round caterpillar tracks, were put into use on the battlefields of Flanders. Aircraft, originally considered to have few military uses, found ready application in surveillance, and airships and balloons, particularly from the German side, expanded from the observation role into a weapon of aerial bombardment not confined to battlefields. Innovation in the war was not confined to engineering developments: chemical weapons had been used before, in the Boer War, for instance, but the First World War saw the range and the scale increased.

New devices for the battlefield: but by the time the Mark I tank was used on the Western Front the war had become firmly stuck in the mud of the trenches

Manoeuvrability and vision had been improved in later First World War tank models, though the real impact of the tank on battlefields had to wait until the Second World War

GREAT BRITAIN

The importance of engineering to the war was not just in innovation, however. Right from the beginning, it was realised that the scale of the conflict demanded the commitment of large sections of manufacturing industry. The production of regular goods was suspended as factories were turned over to war work, producing in huge volumes the consumables of the battlefield – bullets, shells, mines.

This put new strains on industry. It placed manufacturing companies in the unaccustomed position of having to do the bidding of political and military customers. The war work also required high standards of quality and repeatability, the hallmarks of mass production rather than craft-based engineering. And the disappearance into mil-

A new view: airships and aeroplanes brought new dimensions to reconnaissance and to weaponry. This is a First World War convoy in the North Sea as seen from a German Zeppelin

itary service of large numbers of industry's skilled and unskilled workers meant that new workforces, including many women workers, had to be organised and trained.

Early on in the war, there seems to have been a tendency in the Institution of Mechanical Engineers' discussions to regard the war as "an engineers' war" – the true and ultimate test of engineering innovation and production technology. Later, as it became more apparent that the battlefield technology was producing huge numbers of casualties and a tactical stalemate, engineers of all kinds were less keen to identify themselves with the conflict. The realisation grew, too, that the production imperatives of wartime were unlikely to be relaxed with the coming of peace. The war had changed the basics of engineering industry for ever.

For the IMechE itself, the war brought dislocation. A note in the Institution's annual report for 1915 recounts – with the suggestion of a subtle undercurrent of irritation – the gradual commandeering of the headquarters building in Birdcage Walk. This was not the first time the IMechE had let part of the building: the Ministry of Works had been a tenant of some of the upper floors a few years after the first phase was completed in 1899.

Soon after the war began in August 1914, the Institution council offered some space to help the war effort. The Prince of Wales's National Relief Fund moved into the top floor following an invitation from the IMechE council. The government's Office of Works then asked for more space. The council consented "gladly" and the explosives department of the Ministry of Munitions took over the third floor. The explosives department then ousted the relief fund from the top floor in January 1915, took two further small rooms in March and was "loaned" the meeting hall from April.

The report continues: "On 18th May a letter was received from Sir R Sothern Holland stating that it was necessary in the public interest for the remaining rooms in the Institution House to be taken over for the Explosives Department. A Special Council Meeting was held, at which it was decided to point out to the Department that neither the Council nor the Institution, in the absence of special legal authority, had power to hand over the Building, and that such a course would be very inconvenient for the Institution and its members."

In the event, realism prevailed over churlishness. The legal niceties of the individual Institution members being the owners of the building were not pursued, and in June 1915, the Ministry of Works moved in fully and the IMechE moved out.

It did not go far. Temporary offices were found 100 yards away at 11 Great George Street, and the Institution of Civil Engineers, which managed to retain its offices in its own building despite accommodating several smaller government departments, offered its facilities for meetings, as it had done in the years up to 1899. Examinations for membership were held at the Surveyors' Institution, now the Royal Institution of Chartered Surveyors, also in Great George Street. But facilities such as the library were essentially shut for the duration of the war.

Wartime dislocation for the Institution was not restricted to the physical location, though. The Institution finances also took a knock.

The years up to the First World War had been ones of financial suc-

Sir Henry Fowler, locomotive engineer of great distinction and IMechE president in the 1920s, gave a paper on an exhaustive survey of chisels to the wartime Institution

cess. The huge cost of building the headquarters at the end of the previous century on one of the prime pieces of property in Westminster had been taken in the Institution's stride; the extension of the building in 1912 had caused no headaches at all. As late as 1915, the IMechE was reporting income of more than £20,000 and expenditure of less than £12,500, and the total value of the investments and other assets was six times the annual income. It had been a picture of unbroken financial serenity since 1847 – a far cry from the easily won and easily lost fortunes of much of engineering industry in the same period.

Abruptly, all that changed. Members who had volunteered for war service asked for the need to pay subscriptions to be suspended. The Institution noted that other bodies had already taken this step before it agreed. Membership, which had grown steadily in the years before the war, took a turn downwards. There are some indications in the Institution's record that, despite the drive for production, some senior and better qualified engineers found it hard to stay in employment during the war. In 1915, the Institution council considered, and then rejected, the idea that individual mechanical engineers might be allowed to advertise their services. "Such a practice is not in the interests of the Institution, nor of individual members," it said.

At the same that these sources of revenue were being subjected to untoward attrition, costs, partly because of the increasingly successful German blockade, escalated – particularly in printing and paper, where the Institution had committed itself just before the outbreak of war to a regular journal to keep members informed. The result of the loss of income and the increase in costs was that, in the space of just a couple of years, consistent profits turned into consistent deficits.

The wartime restrictions and the temporary loss of the headquarters building meant, in any case, that activities were dramatically scaled down. Annual events such as the Institution dinner, the evening conversazione and the summer meeting in a provincial or foreign city were cancelled. And those meetings that were held tackled subjects which appear to have been almost wilfully uncontroversial.

Following the 1916 annual general meeting, for example, Henry Fowler, chief mechanical engineer of the Midland Railway and later of the London Midland and Scottish Railway, who was IMechE president in 1927 and knighted for his engineering work, gave a long, detailed and technical paper on the subject of chisels.

Odder still, the October 1916 meeting heard a paper from F A Moffatt, described as "of the Argentine Republic". The subject was "Small Machinery in the Argentine Republic". Moffatt's qualifications for such a paper may well have been impeccable, but his range of racial stereotypes, faithfully reproduced in the Proceedings, made strange reading.

The user of machinery in Argentina was, he said, "of the get rich quick type and will buy as a rule the cheap article having the largest output. He is a man who understands little or nothing of what he is buying." Customers were "incapable of judging an article without trying it and greatly carried away by a highly painted and gaudy appearance". Moffatt had as little respect for the country as he had for its engineers: "Practically everyone who goes to the Argentine does

Wartime loss: IMechE president Sir Frederick Donaldson, chief adviser to the Ministry of Munitions, died in the sinking of the Hampshire *in 1916, along with Lord Kitchener*

so with one idea of getting rich quickly and leaving again. The working population to a large extent consists of Italians of little or no education."

It is hard to avoid the conclusion that the learned papers delivered to Institution meetings were almost incidental to the main topic of discussion: the war, which dominated all aspects of everyday life. No Institution member can have remained untouched by the war; for many leading figures in the IMechE, there were poignant memories of the summer meeting held less than a month before the war had started. The meeting's venue had been Paris, but the excursions and factory visits that always formed an important part of both the technical and the social content of the summer meetings had taken the British mechanical engineers to Lille now in the frontline of the war. Many of the factories visited and admired in the summer of 1914 were derelict by the end of that year.

There was poignancy too in the casualty lists. The Institution president at the outbreak of the war had been Sir Frederick Donaldson, the son of the first prime minister of New South Wales, who was chief superintendent of the Royal Ordnance Factories, based at the Arsenal at Woolwich. Naturally, once the international situation worsened, his attention was engaged elsewhere, and vice-presidents filled in for him. "Sir Frederick has not forgotten the Institution," vice-president Michael Longridge told the first meeting after war began, "but the calls upon his time and energies in providing for the needs of the army and navy are now so great that it is absolutely impossible for him to attend to other duties."

Donaldson was destined never to return. Promoted in 1915 to be chief technical adviser to the Ministry of Munitions, which was headed by David Lloyd George, he was one of the senior figures chosen to accompany the Secretary of State for War, Lord Kitchener, on a mission to Russia in June 1916. Their ship, the *Hampshire*, was sunk by a German mine off the Orkneys, and all on board died.

With Kitchener and Donaldson on the mission to Russia was another prominent member of the IMechE, Leslie Stephen Robertson, who had been secretary of the Engineering Standards Committee since 1901. The translation of British Standards specifications into French, Spanish and Russian had been sponsored by the Institution with the aim of helping British engineering trade after the war.

Also a casualty of the war was Professor Bertram Hopkinson, head of the engineering laboratories at Cambridge University, who died in a flying accident while working for the new Air Ministry right at the end of the war. The deaths of famous men in wartime were announced solemnly at the beginning of the next Institution meeting.

Members joined up in their thousands. In the first weeks of the war in 1914, the Institution, along with the Institutions of Civil and Electrical Engineers, sent a circular to members suggesting the formation of an engineers' infantry battalion for service at the front. This idea was rapidly overtaken by an invitation from the Admiralty to the three institutions to steer their members towards an engineer section for the new Naval Brigade. A new circular was sent out, and by the October 1914 meeting, 550 engineers had enlisted in the new unit, "a fair proportion" of them IMechE members, said vice-president Michael Longridge. The councils of the three institutions presented the unit with a set of brass band instruments.

The first Institution member to die in the war was Engineer Lieutenant-Commander Thomas Arnold Venning, who was lost with HMS Pathfinder after a submarine attack off St Abb's Head, east of Edinburgh, on September 5, 1914. At much the same time, graduate member Roland McGroarty, who had delivered a paper to the graduate section a year earlier and who had volunteered as a motorcycle scout at Nairobi, was wounded in fighting in East Africa and died after being captured by the Germans.

The death toll of Institution members rose steadily as the war continued, and in some cases the technological development of the war can be traced in the doleful notices of their deaths in the Proceedings. In 1915, Lieutenant W Inchley of the Duke of Wellington's Regiment was gassed; the following year, Lance-Corporal W A Reynolds of the Royal Warwickshire Regiment was killed by an aerial torpedo; also in 1916, Second Lieutenant G S Hall of the Royal Flying Corps was shot down. And in 1918, Second Lieutenant M S Macaulay of the Lothian & Borders Regiment was killed in an air raid.

By the end of the war, 88 members and one man from the Institution staff were known to have lost their lives. The war memorial listing their names was unveiled at a service held by the Dean of Westminster early in 1921. Three years before, with the fighting showing no signs of respite, the IMechE had joined with other engineering institutions in a service of commemoration for the engineering casualties at Westminster Abbey.

Despite the sacrifice of individual members and the repeated recitation at Institution meetings of the hundreds of members who had joined up, there was a feeling particularly in the early years of the war that the IMechE could and should be doing more. There were worries that too many engineers, as well as craft grade mechanics, had joined regiments that would make insufficient use of their engineering talents.

In addition to manpower concerns, the IMechE also considered whether it ought to formalise its role in disseminating information about innovation. In 1915, it agreed to a request by the Director of Fortifications and Works to supply a list of names of mechanical engineers "with whom the War Office might communicate in connexion with problems arising out of the war".

The same year, a petition signed by 56 members requisitioned a special meeting. The engine manufacturer Ernest Petter proposed a motion to give the IMechE a specific role in channeling innovation through to the war effort. "In view of the highly technical character of the war and the dependence of the Allied forces on the products of mechanical engineering, it is highly desirable that the Institution should in its corporate capacity endeavour to assist the Country by making arrangements for receiving from its members particulars of inventions relating to apparatus likely to be of service in prosecuting the war, improving means of production, or otherwise, and considering, inspecting, reporting upon and, where considered desirable, bringing the same to the notice of the Government."

Petter cited developments such as "submarines and Zeppelins, interchangeable shells, prismatic lenses for periscopes, new explosives, star shells for use at night, new automatic gun sights", all of which depended on "the genius of the mechanical engineer". But the motion did not find favour. A total of 275 members, a high turn-out for any IMechE event, attended the special meeting, but the vote was lost by 66 to 99.

Even without the mandate for an official policy of passing on innovative ideas, though, the IMechE was firmly in the business of looking for useful ideas and passing them on to the War Office. The Institution Journal carried a series of notices inviting members to tackle specific problems: several designs for a mechanical bomb-thrower resulted, and there were proposals "for destroying barbed-wire entanglements, for clearing mines from the products of the explosion of the mine, and for non-slip chains for rubber tyres of motor wagons."

Away from the Institution, of course, many members were caught up in the business of developing military hardware. William Ashbee Tritton, one of the designers of the prototype tank, was an associate member, and another member, Giffard Martel, was appointed brigade-major in one of the new tank regiments in 1916. Martel's later writings on tactics, including the engineering problems of tank-bridging, influenced not just the British army's use of tanks: they also inspired Guderian, the architect of the new German army, whose creation, Rommel's Panzer Division, Martel was to meet when called back into service in 1940.

Lucien Legros' IMechE paper on new methods of achieving traction on bad terrain was subject to censorship: this picture of an Austin experimental vehicle survived the censor's cuts

Reference to military work barely figures in the papers presented to the IMechE during the war. In 1918, the pioneer automotive engineer Lucien Legros described the original caterpillar-tracked tractor in a paper called "Traction on bad roads or land", but the same author had to get official War Office permission as late as 1920 to deliver a paper on tanks to the French Societe des Ingenieurs Civils.

W Cawthorne Unwin, the age-
ing engineering aademic whose
IMechE presidential address
was an early recognition that
British industrial performance
had been less impressive than
Germany's

In the absence of technical detail on innovations, the Institution concerned itself very much in the war years with factory organisation and production matters. This was a natural topic to be covered, since the war made heavy demands on factories. The insatiable need for military hardware and munitions required increases both in production and in the quality of the products, and British factories appear to have felt, at times, that they were shouldering the burden for all of the Allies. Equally, the loss of skilled manpower to the war put new stresses on management skills, including the organisation and the training of a new workforce, mostly unskilled and in many cases female.

But there was a deeper reason why factory and production matters should have bulked large in the IMechE's wartime affairs – and why they should have figured strongly in the Institution president's addresses in the war period. British engineering industry was going through one of the earliest of its periodic crises of confidence about its efficiency and capacity in international terms.

Questions about Britain's international standing and about the country's preparedness for war surfaced in the first presidential address delivered after the start of the war, by the materials specialist W Cawthorne Unwin in April 1915. Unwin, who was already in his late 70s, appears to have had a naturally pessimistic turn of mind. He began his address with a less than jocular reference to his own frailty, appearing to question whether it had been wise for the Institution to elect a president with a lifespan that was likely to be limited. In fact, of course, he lived to be 95, surviving five of his successors.

Unwin's address compared British and German industry and attitudes to engineering in the two countries. The results, he said, were "not entirely favourable" to Britain.

German production of pig-iron, for example, had risen from 11 to 19 million tons a year at the same time that British production rose from 9 million to 11 million tons. German production of steel was twice Britain's. "The German Government is poorer than ours," he said, "but it has much more clearly recognised the interdependence of science and industry, and the duty of the state to assist industry in matters beyond private initiative." Banks were more supportive toward innovation; the Germans had managed to develop world monopolies in some commodities; Germany and Austria sent four or five times as many students to technical high schools as we did.

It was a remarkable speech, and not just for the fact that it introduced themes that would recur across the years to the present day. Unwin was, of course, at pains to point out, several times, that his remarks should not be taken as approval of Germany. "If I speak with admiration of the intellectual efficiency of German education, it is not that I like the Germans or that I do not see some serious evils in their system of education," he said. "It is no condonation of the military crime of Germany to recognise that the enormously rapid industrial advance in that country has serious lessons for us."

But if Unwin was concerned to be anti-German, he seems to have been less bothered about appearing to be anti-Establishment. "I readily admit that the English system of classical education is the best for Civil Service and administrative posts," he said. "I feel sure it is not a good preparation for an engineer." Knowledge of "two dead lan-

Britain's iron and steel industry output by 1915 was dwarfed by Germany's: this is the Newport Iron Works around 1900

guages" was "unsuitable for lads going into practical careers" and "part cause of our discreditable ignorance of continental languages".

Unwin's views led to some disquiet. Unusually, in October 1916, towards the end of his two-year stint as president, he returned for a second presidential address, partly to respond to criticisms of his previous speech, partly, one suspects, to put a rather more positive gloss on his opinions.

A speech at the Royal Society of Arts had questioned some of Unwin's earlier assertions on iron and steel production, on the basis that Germany was a bigger country with a more extensive rail network: that, said Unwin, was to ignore the vast markets for British iron and steel in the Empire, particularly in India. The IMechE president seemed, in any case, to have received the backing of the Prime Minister. Herbert Asquith had spoken of "the leeway we have to make up as regards scientific research and its application to technical and industrial purposes".

Unlike his previous presidential address, though, Unwin was now prepared to admit that progress had been made. "We were unprepared for the war which was forced on us," he said. "But the way in which our characteristic unpreparedness, slackness and inertia have been overcome and organisation created is extraordinary."

He was not, though, going to forsake entirely his tendency to see clouds rather than silver linings. "When the war is over there lies ahead another strenuous time. There will be the cessation of war expenditure, the return of the men at the front with physical and mental capacity enlarged by the experience and discipline of service, and the disposal of the women who, in munition factories and elsewhere, have been replacing men and earning unaccustomed wages."

The twin themes of industry's competitiveness and the coming adjustment to peacetime conditions took up an increasing proportion of the IMechE's time over the years from 1916 onwards. The difficulty of supplying members with technical information on topics which were subject to official secrecy rules was undoubtedly a contributing factor, but this new emphasis on "business" aspects of engineering does appear to represent a conscious shift in the period of the First World War in the role that the Institution perceived for itself. The shift is linked to the earlier adoption of stricter qualification procedures for engineers: from this time onwards, there is a strong undercurrent of opinion that a primary role of the Institution was to help industry to put its affairs in order, and that to do so the Institution's own affairs had to be above reproach.

In practical terms, the shift in emphasis meant a lot more papers on "softer" subjects such as workplace organisation.

It was in this area of organisation and labour that the war was also forcing a shift in preconceptions. Late in 1916, the Institution was trying, without success, to find someone to deliver a paper on the work of women in the munitions factories. There was a lot of novelty in the idea of women in engineering – other industries such as textiles had long relied on a high proportion of women workers, but engineering had been virtually exclusively male.

The Institution itself had followed the wartime trend, replacing some of its absent male employees with female clerks. But there is still a strong suggestion of raised eyebrows in the remarks of IMechE president W Cawthorne Unwin on women in factories: "It is believed that their services have been invaluable, and that for such work as they can undertake they are little inferior to male workers, either as regards quantity or quality."

It took two years before an author for the paper on women workers was found: in 1918, Miss O E Monkhouse of the Home Office became the first woman to speak at an IMechE meeting. The same meeting showed that non-technical topics could be just as much subject to official restriction as technical ones. A companion paper by another

The first-ever paper by a woman, in 1918, was announced in the Journal *on the IMechE's wartime ration of blue-coloured paper*

Michael Longridge, president in 1917, was a continuing presence at the higher levels in the IMechE during a period of considerable change and dislocation

civil servant was hastily withdrawn on government orders, even though it had already been cleared for publication. IMechE president Michael Longridge made little attempt to conceal his annoyance, and the paper was reproduced in the Proceedings and discussed at meetings in Birmingham and Manchester, apparently in defiance of the orders.

Longridge as president continued Unwin's themes of competitiveness and reconstruction and was even more damning in his address, delivered at what was seen by many as the low point of Allied fortunes in the war, about past industrial performance. The mechanical engineer, he said, had been largely responsible for the transformation of Britain from an agricultural, self-supporting nation into "a manufacturing community dependent for its existence upon imported food". The engineer could not therefore escape responsibility for the efficiency of the industries that enabled the food to be bought: he had become "a Trustee for the British Empire".

In this role, Longridge said, engineers had been found wanting. "I think we are willing to confess that our heads were turned by elation at our prosperity, that we were obsessed by admiration of our own achievements; too confident of the sufficiency of our limited knowledge; too contemptuous of the few who tried to throw the light of science on our path; too eager for wealth, and the social influence wealth could buy in the new state of society; too careless of the needs and aspirations of the 'hands' who helped to make the rapid accumulation of large fortunes possible."

It was apocalyptic stuff, and Longridge did not deny himself the logical conclusion: "For every lapse from the ideal, and there is an ideal even of industrial polity, Nemesis Adrasteia sooner or later enacts retribution," he said.

Despite this evidence that he had himself had a classical education, Longridge backed Unwin's analysis that the education system was in need of change, attacking "our two most famous universities" for retaining classical Greek as a compulsory subject. "This reacts upon our Public Schools, and is a serious handicap on those who, intending to deal with the concrete rather than the abstract in their future lives, yet wish to find their level in the social life and moral discipline of

Longridge could trace his engineering lineage back to the Bedlington Iron Works in Northumberland, an early supplier and customer of George Stephenson

these two Universities. The English Public School boy can generally be relied on to face difficulties, lead men, and keep his hands clean in business. Engineering cannot afford to lose him to satisfy the country parsons who rule Oxford and Cambridge."

But it was not just education for the upper classes that did engineering no favours, Longridge said. Apprentices had few opportunities for technical or general education, and British engineering works were generally too small to run their own training schemes. It would be better to raise the school-leaving age to 15 or 16, instead of the present 12 or 14, he thought, and to put the onus for general education on to the state.

Longridge was an industrialist from a family long-steeped in manufacturing: an ancestor had been involved with George Stephenson at the Bedlington Iron Works in Northumberland a century earlier. He was an influential figure in the reorientation of the IMechE into a body more tightly focused on industrial needs, and his presidential address contains the clearest exposition of the new role.

"Mechanical Engineers are broadly divided into two classes, manufacturers and consultants," he said. "The latter, I believe, consider themselves the superior class, because they have been able to impose their wills and specification on the manufacturers. They can do so still to some extent. In the future they will do so not at all." The future for engineers with scientific training was in manufacturing industry, he believed, and that was where the Institution should devote its energies.

Longridge is the consistent figure at the helm of the IMechE in the years from 1914. When Sir Frederick Donaldson became too embroiled in military matters to complete his term as president in 1914, it was Longridge who stepped in; even when the war was over, and Longridge's own term as president was up, he was still in the driving seat. The northern industrialist and newly-elected MP Edward Hopkinson was chosen to be president in 1919, but fell ill before he could deliver his presidential address – in fact, Hopkinson died in 1922 without ever returning to the Institution. Again, Longridge stood in to deliver the president's message.

Hopkinson's address completed the process of refocusing the Institution on industry – at least for the time being. His theme was industrial efficiency: "We have omitted to realise that to make an efficient machine is one thing, but to make it efficiently is another thing. It is no exaggeration to say, for example, that the great majority of British engineering works before the war were utterly unprepared to deal effectively with high class repetitious engineering work. Except for the National Arsenal and a few other engineering works, the system of working accurately to gauges with defined limits and tolerances to ensure interchangeability of parts was almost unknown before the war."

For Hopkinson, though not for some industrialists of the immediate post-war years, there was no going back on the social welfare legislation that the reforming Liberal government of the years immediately before the war had introduced. Measures such as pension, sick pay, holidays with pay and reductions in the working week were cited by many employers as the prime reasons why British industry appeared now to be suffering in comparison with its international competitors;

trade unions, the growth of which had paralleled in some ways the rise of professional bodies such as the IMechE, were also thought by many to be wielding undue influence in the day-to-day affairs of industry.

Hopkinson was having none of this. He quoted from the report of the Industrial Fatigue Research Board: "A workman's output is actually greater when working 48 hours a week than when working 53." He advocated the use of time and motion study and the kind of scientific management techniques pioneered by Frederick Winslow Taylor in the United States to measure and then optimise output. And he insisted that this was the proper concern of engineers: "Management and organisation are vital matters which ought not to be left to chance and should form part of the specific training of a Mechanical Engineer."

This brisk message from a dying man was delivered at the last session of the Institution to be held at the Civils before normal service was resumed. In November 1919, the Ministry of Works finally moved out of Birdcage Walk and the IMechE moved back in. The meetings of the graduate section had started up again the month before; the intention was that the usual range of activities should be back in full swing in 1920.

The war had changed things, though, and not just in the direction of the Institution towards industrial as much as technological innovation. The period immediately after the war was also one of upheaval in Birdcage Walk. The upheaval had its casualties: the long-standing secretary Edgar Worthington had what appeared to be a breakdown early in 1920, and was unable to continue. Worthington had succeeded to the job in 1897 when much the same had happened to his predecessor, Alfred Bache: a breakdown, followed by a long sabbatical, followed eventually by retirement. In the autumn, Worthington was replaced by Magnus Mowat, a former brigadier-general, and the degree to which the administration had expanded is indicated by the fact that Mowat was now to be backed by two assistant secretaries, one dealing with business matters, the other with technical questions.

Upheaval is also indicated by the number of council and committee meetings held in 1920 – a total of 115. One reason for this unprecedented flurry of activity was a series of changes in the basic structure. The changes included the formation of regional branches, new kinds of Institution meeting, and a revision to the ruling council entailing a new class of membership. Finances also came up for debate.

The IMechE had held out against forming regional branches for several years. The reasons for doing this had become less and less convincing over time. Other national engineering institutions had happily created regional operations without seeming to lose cohesion; also the engineering profession appeared to have sorted itself out into an institutional structure which allowed for nationwide professional bodies as well as regional societies requiring fewer qualifications, so the argument that a regional structure for the IMechE would tread on the toes of smaller local societies scarcely held any more; and the frequency with which the question of a regional structure came up indicated that there was a strong demand.

The Institution's position was increasingly anomalous. Papers from

Britain's Empire in India was both a market for British goods and engineers and increasingly a competitor: this is the locally built locomotive used to draw the royal train in the early years of the 20th century

London meetings were often read again at meetings in regional centres organised by honorary local secretaries whose names were mentioned, with grateful thanks, in the annual report each year. Discussion from these regional meetings was reported in the proceedings. But these regional meetings were not accorded branch status. Similarly the Thomas Hawksley lectures which started in 1913 were taken to provincial centres after the first performance in London.

The Institution had in any case managed to get itself into something of a mess with the one branch organisation that it had allowed. In 1908, with the blessing of headquarters, members of the Institution based in the Calcutta area had formed themselves into the first branch. At the same time that proposals to form branches in regional centres in Britain were being frustrated, the Calcutta branch was a source of some pride for the IMechE. Its record of meetings and dinners was faithfully published each year in the annual report, and copies of papers read to meetings in Britain were speedily dispatched to Calcutta to be discussed there. Institution presidents made occasional overseas visits with summer meetings and official delegations in the years to 1914; none went to Calcutta but several managed to slip mention of the branch into their presidential addresses or in the preamble to other meetings. It illustrated the international character of engineering and the extended reach of the Institution itself.

During the war, though, the source of pride became a source of concern. In 1917, the Calcutta branch proposed the formation of an All-India branch of the Institution, with its own offices, secretary and administrative staff. The IMechE council turned this down "for financial and other reasons" and suggested instead that an association of the local sections of the IMechE, the Civils and the Electricals would cost less and could provide the facilities which were evidently wanted. This suggestion seems to have met with no favour in India. Instead, and after some testy to-ing and fro-ing, the Calcutta branch threw its lot in with the newly-formed Indian Institution of Engineers, becoming part of its Bengal association.

This put the IMechE in London in an awkward position. It wanted to support its members in India, but it was not keen to embrace a new body with uncertain qualification procedures as an equal. The 1919 annual general meeting sent a carefully worded telegram, welcoming the formation of the new Indian Institution, but studiously avoiding any offer of reciprocal recognition of qualifications.

At this point, the IMechE was back in its traditional position of having no branches. But a year later, this suddenly changed. All past objections were dropped, by-laws were amended and the Institution announced that local branches would be formed "in a few of the larger centres of Mechanical Engineering". The intention was that the branches should be used for the presentation of papers already presented in London, and for discussion; almost immediately, though, the first branches started organising their own events, sometimes collaborating with local sections of other institutions.

Initially, branches were formed in Birmingham, Manchester and Leeds; these were quickly followed by South Wales and Glasgow, and then by a Western branch based at Bristol. By the end of the 1920s, there was also an East Midlands branch – and a River Plate branch for IMechE members in Brazil, Argentina and Uruguay.

The regional branches soon formed regional graduates' sections as well, and the national graduate members section, which had held its own meetings in London, transformed itself into the London graduates section to match these regional groups.

Meetings in London also changed. The pattern since the Institution was founded in 1847 had been for a number of set-piece papers to be delivered, with a discussion after each. In the very early days, meetings were held quarterly and consisted of one or two papers, and the papers themselves were not long. George Stephenson's paper on rotary engines read to the meeting in 1848 less than a month before his death ran to fewer than 300 words.

The number of papers soon expanded, and a synopsis, rather than the full paper, was usually read to the meeting, often by the secretary, with the author fielding the discussion at the end. By the end of the 19th century, the frequency of meetings had expanded to monthly and some of the papers presented to the IMechE were not much short of textbook size – some take up more than 100 closely-typed pages in the Proceedings. Papers tended to be circulated before delivery to give members the chance to make reasonable contributions to debate at the meetings. Even this was not enough and, by the early years of this century, the discussion was often extended across several months and included written as well as spoken comments. All the while, the intervals between meetings were getting shorter and shorter, with frequent resumptions of discussions held in the weeks between the regular monthly slots.

The idea of informal meetings, which started in 1920, was to allow less constrained discussion of engineering topics, and to promote the more social and less formal aspects of Institution membership. One concession was that members were allowed to smoke in the informal meetings.

Topics for informal meetings varied between technical, even product-oriented, subjects and wider discussions on general industrial issues. In 1921, for example, there were informal meetings on ball and roller bearings and on "our part in the industrial crisis of today". Discussion in informal meetings could also be carried over from one

Captain H Riall Sankey, president in 1920, was in charge at the time the regional branches were set up and was the last of the two-year presidents

meeting to the next.

A further innovation in 1921 was the first joint meeting with another institution: French chemical engineer Paul Kestner presented a paper on the degassing and purification of boiler feed water to a session with the Society of the Chemical Industry's chemical engineering group. The meeting was considered a success, and a further joint session was arranged for the following year.

At the same time that the Institution was changing the way in which it involved ordinary members, the new president, Captain H Riall Sankey, was putting into action his plans to shake up the upper echelons. Sankey, who had taken over as president when it became clear that Hopkinson was too ill to manage the second half of the normal two-year presidency, was concerned that the Institution council, increasingly the decision-making body now that overall membership was heading towards 10,000, was overloaded with distinguished veterans.

He had a point. Past-presidents sat on the council by right, and though Sankey himself had at the most only seven of his predecessors available, and two of those died during his term of office, there was potential for figures from the past to dominate. In addition, the council had six vice-presidents, who comprised the list of names from which future presidents "emerged" for an uncontested election, much in the manner of the old procedures for electing leaders of the Conservative party. In practice, of course, some of the vice-presidents had no real intention of ever becoming president or were unsuited to the role, either temperamentally or because they had too many other commitments. The vice-presidency of the IMechE was as far as they aspired, but, once there, they were difficult to shift. Sankey proposed that vice-presidents should stand down after a maximum of eight years, to become past-vice-presidents.

The new council would then consist of the president, the six vice-presidents, 21 ordinary members and no more than four past-presidents or three past-vice-presidents. The intention, said Sankey, was "to increase the rate of promotion through the council and therefore to be able to get new blood on the council at an earlier date".

All of these changes – the branch structure, the new kinds of meeting and the council changes – were intended to restore something of the immediacy of the early years of the Institution. But there were two other reasons why it was politically expedient for the IMechE to be seen to be taking an active line.

One reason emerged at the annual general meeting in February 1920. An Institution member asked from the floor why there was no mention in the paperwork of engineers' registration. The idea of a register of engineers, which would become one of the continuing themes of the next 75 years, had been raised by the Institution of Civil Engineers. The legal, medical and accountancy professions had managed to gain for their professional bodies the right to restrict the use of the title, and jobs in the public sector at least that required professional qualifications were limited to members of those bodies.

The ICE wanted to achieve the same for civil engineers, and was in the throes of drawing up a register of civil engineers and a parliamentary bill to give legal force to restrictions on the use of the title. The move had obvious implications for the IMechE. Though the

Civils were the senior engineering institution, they had, over the previous 50 years, come to be regarded as representing just part of engineering. No-one seemed clear what effect restricting the use of the title "engineer" in this one part of engineering would have on the other parts.

The IMechE took a firm line. Mark Robinson, one of the perennial vice-presidents who would later become the first past-vice-president, told the 1920 annual general meeting that the council "entirely objected to the Bill as it stood and would oppose it if necessary". Captain Sankey, installed as IMechE president later in the same meeting, was in a more difficult position. "I occupy a somewhat paradoxical position with regard to the Registration Bill," he told the annual dinner at the Connaught Rooms in April 1920. "I am on the committee of the Institution of Civil Engineers which is promoting the Bill and I am on the committee of the Institution of Mechanical Engineers which is opposing the Bill."

Sankey's schizophrenia may well have been instrumental in finding the solution – and in ensuring that relations between the IMechE and the ICE, often tense, did not deteriorate further. In any case, the Civils found that a Private Bill on a matter judged to be of public importance was not allowable, and they were advised that a Public Bill would bring a lot of countervailing arguments out into the open. As the architects had done before them, they quietly abandoned the idea.

This was not the end of the matter, though. Despite the IMechE's objections to the possible hijacking of the title by one part of the profession, Sankey seems to have been correct in identifying that status was a matter of concern to institution members, and that some form of registration might help. "Engineers of all kinds in this country desire a registration of engineers in order that the status of the engineer may be raised to the position it ought to occupy," he told the 1920 annual dinner. "For what can be done without the engineer? Absolutely nothing at all."

It would be many years before the question of status would be tackled, but Sankey's recognition of the desire was popular. The dinner guests cheered him and sang "For He's a Jolly Good Fellow."

So the need to defend the title and to promote engineering was one reason why a higher profile was politically expedient for the IMechE in the early 1920s. The other reason was money.

The IMechE's finances after the war were in the unaccustomed position of being a cause for concern. The Institution's financial record since 1847 had been astonishing. Even in its very first year of existence, with a membership climbing from the original 70 to 103 and revenues of just £515, a profit of more than 40 per cent had resulted. The serene acquisition of profits had not been dented by the employment of permanent staff, the move to London, or the construction of the Institution's headquarters, which was financed by the sale of debentures to wealthier members. On occasion, the accumulated capital was topped up by donations and bequests, but principally the money came from normal day-to-day operations: the Victorian and Edwardian Institution was a steady earner.

It was the First World War that changed this, putting up costs dramatically. The costs of paper and printing for the monthly journal and the Proceedings rose especially sharply. The 1920 annual general meeting was presented with accounts showing a loss and was told that the Institution "like everyone else, had to 'wake up' after the war". Subscriptions, which had been held at £3 a year for members since 1847, were raised to £4, with the fees for associate members also increased. Entrance fees, which had been fixed at £5 in 1847 to include the first year's subscription, were now to be a flat £5, with the subscription to be paid on top of that.

The changes to the finances took some time to produce the desired result. The 1921 annual general meeting was presented with accounts showing a deficit for 1920 of more than £6,000, and questions were asked – a little unfairly, since the subscription increases

Halfway to the 1997 celebrations: the IMechE's 75th anniversary in 1922 was commemorated in the menu card for the annual dinner at the Connaught Rooms

had not come into effect. "The revenue is on a prewar basis, the expenditure is on a postwar basis," the meeting was told. The printing bill in particular had increased by £5,500. There was, though, criticism of some regular items of expenditure: the 1920 annual dinner, the summer meeting at Lincoln and the portrait in oils of the past-president W Cawthorne Unwin had together cost more than £1,000 and were singled out for comment.

Rather conspicuously, the Institution's summer meeting for 1921 was held in and around the headquarters building in London, and took the form of a conference on "means of improving the thermal efficiency of heat power plants". Though there had been collections of papers on adjacent topics given at individual meetings in the past, this was the first specific conference to be run by the IMechE, and it was a big success. It was also timed to coincide with the visit of a delegation of American engineers to Europe, led by the machine tool pioneer Ambrose Swasey.

The saving the temporary change to the summer meeting produced was certainly no more than a gesture, but the mood of the time was that such gestures were needed. It would be a while before the Institution and its members adjusted to the fact that the war, plus the extended range of activities such as publishing, had made economic and financial serenity a thing of the past. Profits were resumed in 1923, but not at the rate of prewar days: from the 1920s onwards, the Institution's fortunes tended to run pretty much in parallel with those of the engineering industry.

The spring-clean of by-laws and bureaucracies continued into 1922. New grades of membership – the Student grade and the Companion – were created, taking the number of classes of member up to seven, in line with other Institutions. New research committees, including a joint venture with the Institution of Naval Architects and the Institute of Marine Engineers on marine oil-engines, were set up. In another deliberate break with tradition, the presidency after Sankey became a one-year post.

In the middle of all this change, the Institution debated whether to apply for a royal charter. Two factors led to the debate. The ICE, which had been since 1828 the only engineering body with a royal charter, was applying for a supplementary charter that would allow its members to call themselves Chartered Civil Engineers: this was seen as a method of achieving the benefits of registration without the drawbacks. But there was a second factor: this time the ICE was not alone. The Institution of Electrical Engineers, which had been blocked by the ICE when it had tried for a royal charter back in the 1880s, was applying again, and this time the ICE was raising no objections.

The IMechE took legal advice on what the change might mean. Somewhat surprisingly it was told that if it attempted to place restrictions on the use of the term "mechanical engineer" or to encourage members to use the chartered mechanical engineer designation, it was likely to face considerable, and possibly justified, opposition from other engineering bodies. Quite why this applied to the IMechE and not to the ICE or IEE is not certain.

The Royal Charter, granted to the Institution in 1930, set the seal on a decade of organisational change

A more cogent reason for rejecting the royal charter at this stage was that chartered status for the Institution meant that all future changes to by-laws and constitutions would have to pass through the Privy Council. Chartered bodies, it was warned, tended to be conservative bodies, unable to change easily. As the IMechE was in the throes of big constitutional changes, and as the overriding criticism of the existing Institution from before the war and after had been its reluctance to change, the council decided not to apply.

The appendix to this, of course, is that at the end of the 1920s, at the time when the next batch of changes in the Institution's structure was going through – including the by-then-welcome abolition of Sankey's creation, the post of past-vice-president – the IMechE decided almost without debate that a royal charter was the way to proceed. From April 22, 1930, when King George V signed the IMechE's charter, all members were able to call themselves Chartered Mechanical Engineers. Many of them had been doing so in any case for years beforehand.

The British Empire Exhibitions of 1924 and 1925 at Wembley had echoes of the Great Exhibition of 1851

The president at the time, Loughnan St Lawrence Pendred, the second of three generations of the Pendred family to edit The Engineer and the second engineering journalist, after William Maw, to be IMechE president, spoke of the royal charter raising the Institution to "a higher dignity" and warned, just a little portentously: "We are now responsible to the King for the maintenance of the fair reputation of our Institution." In practical terms, though, the change in legal status made little difference and when, in the inflationary 1970s, changes to the charter seemed to be needed only too frequently, the in-built resistance to change from the Privy Council that had been forecast was found not to exist.

All of the changes in the 1920s were designed to meet the criticisms of the Institution that had surfaced before the war and during it, and to address the weaknesses that were perceived in British manufacturing industry and in engineers' contribution to it. The effect of the changes was very much to strengthen the IMechE's role in engineering education, to bring the Institution's technical and learned society function more closely into line with industrial issues of organisation and production, and to widen its national role on official and industrial bodies.

In education, the range of organisations where the Institution was influential grew markedly in the inter-war years. By 1926, the IMechE was represented on the university courts at Bristol, Liverpool and Sheffield and on the governing boards of Imperial College in London and the School of Mines in Cornwall; it was also on the London committee for the University of Hong Kong. One side-effect of the involvement of Institutions, including the IMechE, with engineering courses and universities was to accentuate the differences between branches of the profession, a process that was also affected by the qualifying examinations for membership.

Official bodies where the IMechE was an active participant by the mid 1920s included the National Physical Laboratory, the general committee of the Royal Society responsible for administering government grants for scientific investigation – a forerunner of the research councils – and committees of the British Engineering Standards Association. Wider industrial involvement included participation in exhibitions such as the British Empire Exhibition at Wembley in 1924 and regular exhibitions on smoke abatement, and contributions to an increasing number of conferences and congresses, both in Britain and around the world.

A few of these activities went back to the period before the war. The National Physical Laboratory, for instance, had long been the venue for some of the Institution's sponsored research projects, and the IMechE had also been involved in standards over a long period. But the 1920s saw a dramatic growth in these official and semi-official activities, and the Institution nationally seemed to take a deliberately outward-looking line. Participation in many of these activities would help to put the Institution into a nationally influential position in the Second World War and afterwards, far more than it had been in the First World War.

The inter-war years were also a period of sustained growth in terms of membership. Perhaps inevitably, the vigour of the Institution's new regional sections tended increasingly to overshadow the more laboured, and sometimes rather more anodyne, pronouncements that came from the centre. Many of the best papers reproduced in the Proceedings in the inter-war years come from regional meetings, often dealing with national issues from the perspective of the industry in one area.

Industry and the technology it used changed enormously in the inter-war years. The motor industry and, later, aircraft manufacture emerged as new and important sectors and motor manufacturers in

*Ford brought the techniques of
mass production to Britain just
before the 1914-1918 war, with
its works at Trafford Park, producing the Model T*

As with early mechanical engineering in the first half of the 19th century, early mass production married craft skills with engineering methods. This is Austin's Longbridge factory in the 1920s

Flow-line 1920s-style: chassis assembly of Bullnose Morris cars

particular brought with them the ideas of mass production, some of which the IMechE members on the summer visit to the United States had noted as far back as 1904.

The new production technologies brought with them new demands on mechanical engineering skills; they also brought debate within the Institution about the degree to which engineering workers could, or should, be made subservient to an industrial culture in which the machines set the pace and the parameters. These philosophical debates often ran parallel to practical discussions on issues such as productivity, piecework and pay rates which were at the forefront of the great industrial relations disputes of the 1920s. The IMechE's new focus on organisation and management meant that it did not shy away from controversial topics, as it might have done a generation earlier.

The techniques of mass production also moved out from the motor industry into sectors such as food, which was increasingly organised on a factory system and was also increasingly mechanised; the same

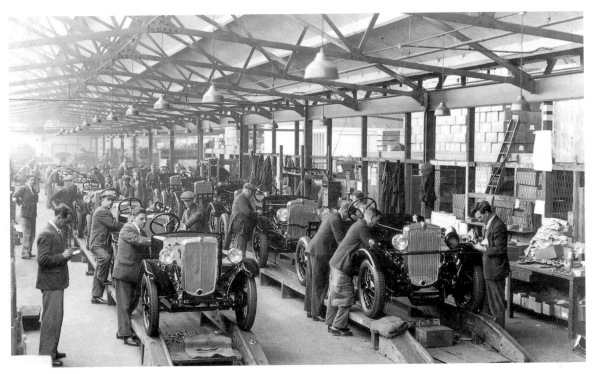

Parallel production lines: Morris Minor and Oxford cars at Cowley in 1933

period saw petrochemicals develop as an industry in its own right. The manufacture of food and chemicals stimulated developments in control systems and process engineering, and these in turn fed back into more conventional mechanical engineering in areas such as machine tool operation.

The depression of the 1930s and the Second World War that followed meant that progress in these areas was often stuttering; the great developments in technologies such as control theory and man-made materials would find application mostly after the war.

But already in the 1920s and 1930s there were changes visible which were akin to those of the Industrial Revolution a century before. There were self-fuelling circularities of development in one area leading to development elsewhere: the motor car drove the

The very beginnings of motor industry automation, with conveyor systems bringing in bodies for Morris Eights in 1934

New industries such as petroleum refining and petrochemicals were assisted by the increasing numbers of cars on British roads

petrochemical industry, which created the plastics, which would, in time, find mass markets in the motor industry. The beginnings of a mechanical engineering economy centred on the internal combustion engine had parallels with the economy of the mid 19th century which was founded on the steam engine, in that it changed industrial geography, encouraged company specialisation and developed interdependencies between component suppliers and the assemblers of finished products.

At the same time that these new industries were becoming established in Britain, existing engineering sectors were expanding or reaching new peaks of achievement.

Engineering technology was pushed towards new limits in many areas: rail, road and air speed records were broken regularly in this period; ocean liners, naval warships and power stations became ever bigger.

A second IMechE president, Sir William Stanier of the London Midland and Scottish Railway, turned to streamlining after the series of engines that started with the Princess Royal. Stanier also experimented with gas turbines in locomotive number 6202 of this class

Again, there was cross-fertilisation of ideas between different mechanical engineering sectors. Work in the aircraft industry on aerodynamics had effects in the motor industry and in railway locomotives: the hallmark of steam locomotive design at this time was streamlining, pursued particularly by three engineers who became

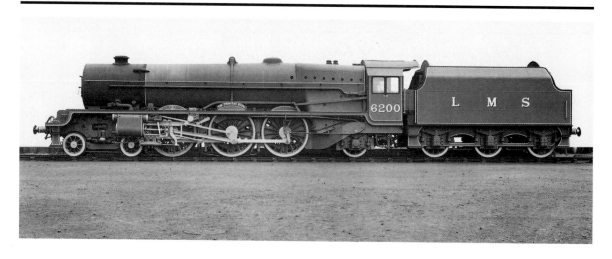

Streamlined for speed: the A4 Pacifics were designed by IMechE president Sir Nigel Gresley. The locomotive on the left bears Gresley's own name; the one on the right is the world steam engine speed record-holder, Mallard

president of the IMechE, Sir Nigel Gresley, Sir William Stanier and O V S Bulleid. Industrial design was emerging as a branch of the discipline of mechanical engineering over and above the basic requirement of drafting as a means of communicating ideas and dimensions.

Elsewhere in the inter-war years there was considerable progress in engineering's contribution to living standards. Many of the final gaps in the infrastructures of electricity, water, gas and roads were completed. London's tube network was extended and electrification was applied to surface railways over a wide area of commuter lines around bigger cities. The first stages of a new and wider infrastructure in air travel were set down with the development of the first airports.

Gas production in the 1930s with manual stoking of horizontal retorts

The gasman cometh: the mains come to Cobham in Surrey. This picture is not dated and could be from any period between 1925 and 1960

The Great Northern, Piccadilly and Brompton railway, now the Piccadilly line, predates the inter-war years; this photograph postdates that period. Yet the London Underground, from its trains to its station signs, is very much a creation of the 1920s and 1930s

Left: Hoover's very first cleaner, the model O from 1908. This model and its successors contributed to a revolution in domestic life in the inter-war period

Right: luxury and convenience in the 1927 Frigidaire refrigerator

There is a strong sense in the work of many of the IMechE branches in this period that mechanical engineering is a very wide discipline indeed. Lists of branch visits indicate a liberal definition – and sometimes a healthy social element to the proceedings.

Even at a national level, where there seemed to be fewer and fewer events as the regions assumed a bigger role, the IMechE took a refreshingly broad view of its remit. Some of the engineering developments of the inter-war years, such as radio and talking pictures, were very dubiously mechanical: the 25th Thomas Hawksley lecture in 1938, though, was on the subject of "Television" and was given by the BBC's engineering director, Sir Noel Ashbridge.

The use of the Thomas Hawksley lecture as an "event" in its own right, not always directly attached to mechanical engineering, had been established several years earlier. Lecturers included the Nobel Laureate Sir William Bragg and the astronomer and mathematician A S Eddington, the physicist whose experiments in the years after the First World War went a long way towards confirming Einstein's theory of relativity. Eddington talked to the IMechE about "Engineering Principles in the Machinery of the Stars", and later Lord Rutherford, the first scientist to postulate the structure of the atom, spoke on his work at the Cavendish Laboratory in Cambridge.

That the IMechE was able to attract these top scientific names says much for its prestige in the inter-war years – as well as for the breadth of its vision. Britain appeared to be very much at this time in the vanguard of scientific and technical development worldwide. Engineering excellence could be measured by the number of British record-holders. One of our locomotives, Sir Nigel Gresley's A4 class Mallard, ran faster than any other; one of our ocean liners, the Queen Mary, and then the Queen Elizabeth, was bigger than any other vessel; Britons regularly increased the world land-speed record across the 1930s.

Record-breaker: Rolls-Royce's Eagle aero-engine powered the Vimy which was piloted by Alcock and Brown in the first transatlantic crossing in 1919

Yet for all the undeniable engineering advances in the years between the two world wars, the dominant factor of the period was not the technological development. Instead it was the faltering and sometimes brutal progress towards a new world economic order based on industrial muscle rather than territorial possessions.

There is an anomaly in the inter-war IMechE. The huge amount of restructuring of the years immediately after the First World War was intended to make the Institution more relevant industrially. The IMechE presidents of the period from 1914 onwards were far-sighted in their identification of the ingredients that would make for industrial competitiveness and frank in their analysis of Britain's shortcomings. The reforms that they put in place made the IMechE at national and local level the forum for discussion of issues such as factory organisation and engineering management.

Yet though the Institution was a body of undoubted prestige in these years – and though membership grew strongly and the restructuring process was universally considered a success – the IMechE addressed only sporadically the wider economic issues affecting its members individually and the companies that many of them worked for.

Industrial change: fewer than 20 years elapsed between the arrival of Ford at Trafford Park and the last vehicle off the assembly line there, in 1931, when the company relocated its main British factory to Dagenham

Even the great depression from the end of the 1920s onwards did not see the IMechE stir itself into more than intermittent action. In 1933, the council formed a committee to consider what might be done about unemployment in the engineering industry, and the president, William Taylor, wrote to the Prime Minister Ramsay McDonald – later an honorary member – offering help with any public works programme. The committee, styling itself rather grandly the Works of National Importance Committee, decided that extending water and gas supplies throughout the country was the kind of project that could help alleviate the loss of jobs. Further letters passed between the IMechE and ministers, but the government was not in favour of a concerted programme of public works along the lines of the New Deal in the United States, and the committee seems quietly to have disbanded.

In other parts of the Institution's record in these years, there is a strong suggestion of nostalgic torpor at the centre, contrasting with the vigour of the regional branches.

The summer meetings resumed in 1922, and that year's meeting was an exercise in international goodwill. The IMechE returned to Paris, where the 1914 summer meeting had been held, and then proceeded to Belgium, where there was an international engineering exhibition. In both countries, solemn ceremonies commemorating the dead of the war were held.

Inter-war engineering was about pushing the frontiers of performance. The Supermarine, the forerunner to the Spitfire; the Miss England III *speedboat; and the* Queen Mary *passenger liner were all record-breakers*

Thereafter, though, the summer meetings seem to have become more and more one of the social events of the Institution calendar: by 1930, when the meeting was held in Bristol, only one paper was read – a review of the past 10 years of aero-engine development – while the number of receptions had increased to four and excursions to engineering works and to tourist attractions such as Longleat had mushroomed.

Attendance at the 1930 summer meeting totalled 288 members and visitors and 139 ladies: the presumption was still that ladies would not be members. A fair number of them took the one-day excursion to Petters at Yeovil

In 1932, the summer meeting was even less rigorous. "Members and ladies" sailed on August 19 for Quebec and Montreal, then proceeded at a leisurely pace through Ottawa, Toronto, Niagara Falls, and Schenectady to New York, arriving back in London on September 19. The 1932 annual report tells the full story of the hard work this meeting entailed: "The business meeting was held in the University of Toronto on 31st August, when a paper on 'The Generation and Distribution of Power in the Province of Ontario with Special Reference to the Work of the Hydro-Electric Power Commission of Ontario' by F A Gaby DSc was read and discussed." And that was that.

The real engineering work of the Institution was increasingly done at regional and local level, and the Proceedings reflected that. In 1933, eight papers from national meetings, against 21 from local branches, made it to publication. Branches held many more meetings than were held at headquarters and organised visits and social activities too. It is hard to escape the conclusion that the national organisation of the IMechE had, by the early 1930s, yielded its original role to the branches and had yet to find itself another role, except the necessary job of representation on other bodies.

Petters' certificate of engineering apprenticeship, collected on the 1930 visit

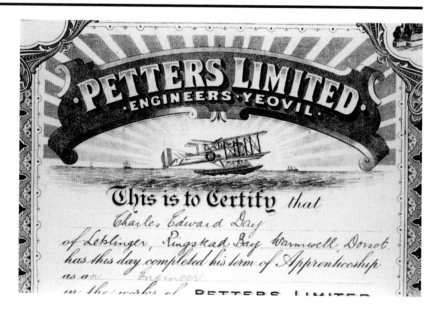

Three factors seem to have changed matters and to have helped to shake the national Institution out of its lethargy from the mid 1930s onwards. The first was a small announcement in 1933. The council appointed a standing committee called the Inventions Advisory Committee to help members with mechanical engineering inventions. The commitment to promote invention had, of course, been part of the original text that summoned mechanical engineers to form the Institution in 1847; progress through invention had been the Victorian ideal which had somehow become lost in the 20th century.

In its first year, the committee advised 25 IMechE members on inventions. There is no evidence that its advice produced any winning inventions and the committee was fairly soon amalgamated with the longer-established Research Advisory Committee – and then both failed to survive the Second World War.

Nevertheless, from the mid 1930s onwards, invention and innovation came back on to the IMechE's agenda nationally and regionally, with the strong implication that through invention might be found the means to overcome the economic tribulations of the depression. There were lectures and discussion meetings on the nature of invention in several regions in the years leading up to the Second World War; in 1934, the Institution nationally held a symposium on invention.

Not all serious purpose: graduate sections of the institutions held events whose emphasis was social as much as educational. This is the Institution of Automobile Engineers' graduate rally – or "Grally" – in 1934

The trend towards holding symposia and conferences was a second factor in the revival of the Institution as a national body after the fallow years. The trend started in a small way from the inventions symposium of 1934 and had barely built up any momentum by the outbreak of war, though the big international conference on lubrication in 1937 is seen as a benchmark event in the history of that branch of engineering. But the success of the growing number of conferences and the collaboration with other organisations, both in Britain and around the world, was a pointer to the development of the Institution's role in the future.

The third change in the Institution nationally came in the 1930s with the setting up of the first discussion groups, the forerunners of the special interest groups that have, in different forms, been increasingly the focus for activity ever since. The first, the Internal Combustion Engine Group, was formed in December 1934. By 1939, it had been joined by four others, covering steam, education, hydraulics and manufacturing, though the last two of these had not in fact held any meetings when the Second World War broke out.

These factors – the restoration of the concept of the IMechE as a focus for innovation, the growth of conferences and collaborations with other bodies, and the establishment of nationwide special interest groups – contributed to a noticeable revival of the Institution at national level in the years just before the Second World War. The calendar of events at Birdcage Walk started to expand again; attendances at London meetings improved.

The Second World War was the culminating factor in this national revival for the IMechE. Unlike in 1914, when the Institution itself and the engineering industry where so many of its members worked were largely unprepared for war, the Institution in 1939 was ready to play a prominent part.

5   War and Peace

1939 TO 1960

M ore than in the First World War a generation earlier, the decisive role in the Second World War was played by machine power rather than manpower. Aircraft, submarines and tanks, which had been used for the first time in the battles of the First World War, were the primary offensive weapons between 1939 and 1945; and they were joined by newer technologies unthought of in the earlier conflict.

Unmanned rockets as well as manned planes in the wartime skies brought aerial bombardment regularly to British cities and put civilian populations in the front line alongside the fighting troops; radar enabled sailors and air force personnel to detect the presence of

The new technology of warfare: the German V2 rockets that brought terror to British cities in the final year of the Second World War

Fig. 1. The German "A4" (V2) Long-Range Rocket

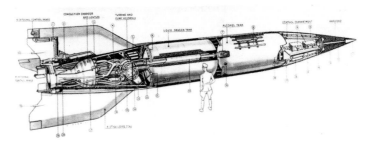

The unsung technology of warfare: simple landing craft revolutionised the concept of amphibious troops and allowed the Allies to begin the long slow process of recapturing France and Italy

unseen enemies; the jet engine promised a new dimension in aircraft speed and efficiency; and the war was finally ended by the detonation of two bombs of a kind which had scarcely been imagined outside academic scientific circles when the war had begun six years before. Engineering and technical innovation made the headlines of both victory and defeat possible.

But engineering was also at work in less glamorous, or less publicised, war efforts: in the development of the humdrum gauche-looking amphibious warfare ships which made possible the heroism of the allied landings in Normandy and in Italy; in the secret work of the code-breakers which helped to set engineering and mathematical minds thinking about the fundamentals of computing; in the clumsy

Radar, developed by the British engineer Sir Robert Watson-Watt, gave the armed services of the Second World War new powers to detect enemy activity

but effective concrete and steel floating Mulberry harbours hastily assembled in engineering companies and shipyards and towed out to war.

In its early stages, the First World War had been widely referred to, not least within the Institution of Mechanical Engineers, as "an engineer's war". No-one spoke of the Second World War in the same terms, yet the description would have been more accurate. The engineers of the earlier conflict had supplied heavyweight immobile armaments capable of inflicting immense damage but incapable of the factor of surprise necessary to give advantage to one or other side. The result was that the frontline of the First World War was a static and closely defined place of trenches and carnage: a battlefield like no other, before or since.

By the Second World War, developments in vehicles, aircraft, rockets and bombs had restored the capability of surprise, and the frontline was Coventry, Dresden, Arnhem, Burma, and, finally, Hiroshima – the technology turned civilian populations across the world into participants, and encouraged new battlefield tactics based on mobility rather than attrition. It made for no less carnage and much more

The Lancaster bomber, powered by Rolls-Royce Merlin engines

Merlin engines, made by Rolls-Royce for wartime bombers

widespread terror, and in the bombs dropped ·on Japan which brought the final surrender in the war in the Pacific, the ultimate capability of technology to be used for destruction as well as creativity was demonstrated for the first time.

Engineering had made possible the stalemate of the First World War; in the generation since, progress in engineering technology had made available the military equipment that enabled the stalemate to be broken. The price of this progress, though, was a heavy one. Engineering, and mechanical engineering in particular, had since the beginning of history been the most potent factor in the construction and development of civilisation. Now its capacity to destroy civilisation as well as to build it stood revealed, not just in the power of the nuclear bomb, but in the perversion of technology to sadistic ends in the holocaust.

Compared with their First World War predecessors, the tanks of the Second World War were manoeuvrable and utilitarian – and a familiar sight on and off the battlefield

Unlike in 1914, when war arrived suddenly only a month after the IMechE's summer meeting had taken in visits to factories in what would become the battle zone, Britain in 1939 had been self-evidently on the slippery slope towards war for the previous year or so, and the Institution had been involved in the preparations. It had been co-operating with the government since the Munich crisis of 1938 in setting up a Register for National Service with several thousand names. The institutional machinery swung into action directly war was declared, and its more important archives and much of the administration were dispatched to the Surrey countryside at Betchworth in the North Downs.

Virtually all meetings, local and national, were cancelled. In fact, because of the worsening international situation through 1939, one of the main events of the session had already been sacrificed. The plan for the year's summer meeting had been for a long excursion to the United States and Canada along the lines of the 1932 extravaganza. Trailers in the Institution's journal advertised a five-and-a-half week visit, taking in a visit to the New York World's Fair. The IMechE party was due to arrive on American soil on Sunday September 3, 1939. But by the time that day came, the IMechE and the rest of the world had other concerns. It was the day war between Britain and Germany was declared.

The American expedition was destined never to be revived, but much of the other early response of the IMechE to war was gradually reversed. Like the children evacuated from London to escape the blitz, many of the IMechE staff sent to Betchworth felt homesick and remote from the action, and contrived to make their way back to the capital within a few months. When, 30 years later, the Institution again proposed moving some staff out of expensive London property to a provincial location, one of the Betchworth evacuees reappeared at a special meeting of the Institution to cast doubt on the wisdom of the move in terms of staff morale. In the event, the peacetime attractions of an out-of-town location in the 1970s proved even stronger for many IMechE staff, and the move was a success.

In the Second World War, as in 1914, parts of the Institution headquarters in Birdcage Walk, a stone's throw from Parliament and the seat of Government, were commandeered by the Ministry of Works and used by a succession of government departments. Unlike in 1914, though, the Institution was able to stay in its own building alongside the government bodies. The evacuation of whole branches of the government from London because of the threat of air raids meant that there was much less pressure on office space close to the seat of power in 1939.

The solidity of the IMechE building was officially recognised: Westminster City Council scheduled the lower meeting hall as a public air raid shelter available day and night for 210 people. But despite its closeness to prime targets for German bombers, the headquarters suffered little damage in the war. An incendiary bomb started a small fire on the library roof, which Institution staff had extinguished by the time the firefighters arrived, and windows were broken regularly by nearby bombs. But the fabric emerged unscathed.

The increased number of members of the Institution in the Second World War meant an increased number of casualties. But as a engineering organisation, the Institution itself came through the Second World War with its reputation positively enhanced.

In contrast to the First World War, the IMechE took a fairly robust attitude towards maintaining its activities and keeping its doors open. It continued and increased the new activities such as conferences and specialist groups that had started to revive its fortunes nationally in the years running up to 1939. And above all, thanks in great measure to the energies of one man, it acquired for itself as a body a central role in national life that reflected properly the contribution that mechanical engineering inventions and products were making to the military effort.

The robust attitude towards the hostilities took a while to develop. In the early months of the war, the policy of cancelling events, at both national and local level, was kept to. Bruce Ball, the president at the outbreak of war, had his presidential address deferred and his second chance to talk to the Institution, at the annual dinner, was then swiftly cancelled. While it was unclear whether meetings should go ahead or not, the Institution appealed to its members to continue to submit papers for publication, though restrictions on paper use eventually took its toll on both the Proceedings and the journal.

Gradually, though – and mostly because there seemed no good reason not to – many of the Institution's activities were revived and Ball was at last able to deliver his address, four months late, in February 1940. Evening meetings remained rarities for the duration of the war, but there were increasing numbers of afternoon meetings, both in London and outside in the regions.

Some of the other activities continued first in a spirit of puzzlement as the expected deprivations of war initially failed to materialise and then as part of a collective mood of defiance as Britain stood alone in Europe. The Dunkirk spirit was certainly evident in the second 1940 batch of examinations for admission to the Institution in October, held in London at the height of the air raids and the Battle of Britain.

Business as usual: incarceration in German Prisoner of War camps did not stop mechanical engineers from sitting – and passing – IMechE examinations. Papers were sent via the Red Cross

PART RESULTS OF OCTOBER 1943 EXAMINATION in 4 German Prisoner of War Camps.

ASSOCIATE MEMBERSHIP EXAMIN...

CANDIDATE'S NAME	NUMBER	GRADE OF MEMBERSHIP	Section A Part I — Applied Mathematics	Physics and Chemistry	Part II — Modern Language	Essay & Fundas. of Indus. Admin.	Section B — Theory of Machs. and Mach. Design	Properties and Strength of Mats.	Steam and the Steam Engine	Internal-Comb. Engines	Hydraulics and Hydr. Machinery	Electrotechnics, etc.	Metallurgy
							Passed Sections B and C						
MORCOMBE,Lieut.V.J.820 Oflag V11B							56	54		84			
							Passed Sections A Part 11 and B						
DAIN, S/Sgt.C.,95375(2823) Stalag 383					F		61	63	60				
							Passed Section A						
RICHARDSON,M.S.M.(R.E.8) V.Q.1873(3482) Stalag 383			56	45	50/60								
							Passed Section A Part 11						
BUCHANAN,S/Sgt.C.G. No.21287 Stalag 383				(h.p. Oct 42)	75/65						(h.p. Apr.1943)		
							Passed Section A Part 1						
TERRY,Lieut.J.F.G.3469 Oflag V11B			35	37									
							Passed Section B						
COVENTRY,W.L.87914 Marlag and Milag Nord.							46	54	83				
MAY, Sgt.A. 2024 Stalag 383				(h.p.)	(p.p) Apr.1943		36	32		44			

But there was even more of a determined normality about the way the IMechE managed to hold examinations at 20 other centres in Britain and Ireland, six in India, four in New Zealand, three in South Africa and one each in Argentina, Sri Lanka, Malaysia, Nigeria and Canada in the same year. And engineers on active service were not missed out: examinations were held on two Royal Navy ships and at three centres "in the field" for the British Expeditionary Force in France in April 1940.

Even this defiant determination that the business of the Institution should go on was topped later in the war: from 1942, using the International Red Cross as an intermediary, the IMechE arranged for its examinations to be sat by prisoners held captive by the Germans.

As the war progressed, the IMechE became bolder about reverting to its normal programmes. In 1942, it organised several joint meetings with the Institution of Electrical Engineers and one conference, on statistical control methods, run in conjunction with both the IEE and the Civils, attracted an audience of more than 1,000. A new specialist group on applied mechanics was started in the same year. Attendances at all events throughout the war were satisfactory, and informal meetings were often packed.

In general the subject matter for wartime papers trod a careful line between the need for security and the evident desire of members for information. This line avoided the anodyne banalities of some of the First World War efforts, but obviously detail on the engineering developments that many members of the IMechE were intimately involved in during the war was subject to official secrecy and had to wait until hostilities ended.

Where the Institution was able to make a particular wartime contribution, though, was in disseminating information about manufacturing technology. The specialist manufacture group, which had been set up just before the war but had yet to hold any meetings of its own, was "attached" to the North West region and a series of well-attended meetings ensued, dealing less with the leading edge of production technology and more with the day-to-day difficulties of manufacturing in wartime and with basic machine tool operation. As many of the machine tools in use in British factories in the war were unfamiliar machines that had arrived courtesy of the Americans under the lease-lend scheme, this kind of instruction was invaluable, and the meetings appear, in any case, to have been kept as informal as possible to encourage questions and problems to be raised.

The work of the manufacture group was part of a wider attempt by the Institution to take a more active role in the war than it had done between 1914 and 1918. The IMechE appears to have been clear that its main role in the war was through professional engineers working in industry: in 1940, it noted that the majority of its members were working in industry on armaments production, though 16 per cent were reported to be in the forces or working for government departments.

Overall, the importance of manufacturing industry to the war effort was better appreciated the second time around, though Britain's national preparedness, in terms of armaments and other war supplies, was not much better than it had been in 1914. The long-standing policy of appeasement of the fascist dictators in the mid 1930s

Much of the British industry was turned over to the manufacture of wartime supplies: this is the Austin factory at Longbridge, where the Stirling bomber was built

Series production at MG's headquarters at Abingdon

meant that military supplies were not built up to the extent that they were in some of countries Britain was not fighting. Industry was, though, quicker to organise itself for wartime production from 1939 because it had had the earlier experience, and because the techniques of mass production had become established in many civilian industries in the inter-war years.

The demands of war had dramatic effects on industry itself, which had barely begun to recover from the depredations of the 1930s depression. Employment in the industry doubled from fewer than 1.5 million to just under 3 million in the first four years of the war. The new workers were needed to produce material for the war effort and as in 1914 the production of many non-military engineering products was suspended.

The British motor industry was one of several sectors that effectively closed for the duration of the war; just 16,000 cars were produced between 1940 and 1944 and factories were diverted into manufacturing armaments, aircraft and armoured vehicles. Morris's Cowley factory, for example, was the only civilian factory in the country allowed to repair damaged warplanes; it also produced mines, which it had made during the First World War as well.

One consequence of the importance of engineering industry to the war effort was that factories were identified as legitimate targets for enemy bombers, and the same raids that exacted a terrible toll on Coventry, its inhabitants and its cathedral also destroyed large sections of the city's engineering industry. The destruction of other industrial cities, especially those with ports, constantly disrupted production. The IMechE, departing from its usual topics of discussion, ran papers on air raid precautions and measures to make industrial buildings more blast-proof.

With many skilled engineering workers as well as engineering professionals in the armed services, many parts of the industry began again to recruit the hitherto under-used segment of the labour mar-

Many IMechE members saw active war service as engineers. Member Major Douglas Henderson, right, recorded locomotive casualties in the North African campaign

ket – women – to maintain and increase production. As in the First World War, the new workforce was relatively unskilled and this factor, combined with the urgent need for maximum production and productivity, increased the mechanisation of industry and the need for organisation in workplaces.

On these subjects, the IMechE was better equipped in the Second World War to make a contribution than it had been in the First. But a lot of the impetus for the involvement of the Institution in issues of national importance came from the work of one man, Henry Guy.

Guy was a Welsh-born engineer of some academic distinction who had become, as chief mechanical engineer of the Metropolitan-Vickers company since 1918, increasingly influential in the IMechE. In 1927, he was awarded the Institution's Thomas Hawksley Gold Medal for a paper on steam cycles, and later he worked with future IMechE president William Stanier on steam turbine driven locomotives. By the mid 1930s, he was a vice-president of the Institution, and in 1936 he was also elected a fellow of the Royal Society, serving on its

Wartime dynamo: Henry Guy, secretary of the IMechE from 1942, took the Institution into the heart of the war effort

council in the years immediately before the war. He was therefore both powerful and well-connected; he appears also to have been a man of considerable energy and pugnacity, who liked to get things done.

The IMechE had gone through something of a hiatus at the top of its permanent staff in the late 1930s and into the first years of the war. The secretary since 1920, Brigadier-General Magnus Mowat, had unfortunately followed the precedent of his two predecessors and succumbed to illness in 1937, finally tendering his resignation in 1938. One of the two assistant secretaries brought in to help Mowat in 1921, J E Montgomrey, took over as secretary, but he was due to retire in 1941. Into the void stepped the figure of Henry Guy.

From the beginning, Guy was much more than just the head of the Institution's administration, and though he took the title of secretary from 1941, he seems to have relied heavily on his assistants and even on Mowat, who was accorded the title of "secretary emeritus", for much of the day-to-day work. He had something of a reputation as a martinet: after the war, as IMechE secretary, he rigged up a system of lights for council meetings along the lines of those used at political party conferences, and ruthlessly operated them to indicate when council members had reached the end of what he considered a reasonable time for making a point.

Even before he joined the staff, Guy had started involving himself – and by implication the Institution – in official work. His list of credits is testimony to incredible energy and appetite for work.

In 1940, he had become chairman of the Royal Society's engineering sciences sectional committee and the following year he joined the executive committee of the National Physical Laboratory. Right from

147

the outbreak of war he had been on the scientific advisory council of the Ministry of Supply; from 1942 he was on the mechanical engineering advisory committee of the Ministry of Labour, and two years later he joined the advisory committee of the Department of Scientific and Industrial Research; he also chaired various official committees dealing with gun design, armaments development, static detonation, the work of the Royal Aircraft Establishment, the technical organisation of the army and aircraft arms research. His industrial work included chairmanship of the British Electrical and Allied Industries Research Association and of several committees of the British Standards Institution.

All of these connections had the effect of propelling the IMechE forward to a central role in the national framework for scientific and technical advice in the war years – and beyond, because Guy continued to play an active role in government-backed bodies when the war was over. The Institution had been overshadowed by the Royal Society and other bodies in the First World War; in the Second World War, very much because of Guy, it took the leading role.

One particular area in which the Institution was able to be of material help to the war effort was in the formation of the Royal Corps of Electrical and Mechanical Engineers, later REME. Guy was again prominent in the discussions that led to the setting up of this separate army engineering corps in 1942, and the Institution helped find officers with specialised engineering knowledge for REME, as it had done for the Ordnance Corps since the start of the war.

The formation of REME and the wartime need for qualified engineers brought a surge in membership in the middle years of the war. There was a record increase in members in 1942, and that record was immediately overtaken by the figures for 1943. The Institution to which engineers returned as civilians in 1945 was transformed from the prewar body.

Industry too had been revitalised away from the 1930s depression by the war and had also been changed – irreversibly this time – by developments in machinery and in workplace conditions.

Above all, though, engineers returning to peacetime employment had much to discuss in the way of technical developments. Few details about military machinery and about the innovations of individual mechanical engineers had been released during the course of the war, and the Institution of Mechanical Engineers moved quickly to satisfy a real need for information.

War-related lectures given in the first few months after the end of the conflict included a Thomas Hawksley lecture by Sir Edward Appleton on the scientist in wartime, which ranged widely from the atomic bomb, through an early description of radar, through to developments in armaments. Appleton, the permanent secretary at the Department of Scientific and Industrial Research, explained that his definition of scientists included mechanical engineers. He said that on the collaboration between scientists, industry and the services "the whole future defence of this country depends".

Specific engineering development of wartime were covered in a talk by Sir Claude Gibb of Parsons on tanks, and one by A C Hartley on Pluto and Fido, the wartime underwater pipeline and fog dispersal projects.

A C Hartley, IMechE president and inventor of two significant wartime developments, the PLUTO pipeline and the FIDO fog dispersal equipment

Hartley, later to be president of the Institution, is a 20th century rival for any of the great names of mechanical engineering invention of the 19th century; he is also an example of the lack of correlation between academic achievement and inventiveness, having managed the dubious distinction of a third class engineering degree. He had been chief engineer of the Anglo-Persian Oil Company before the war and had then been seconded to the Ministry of Aircraft Production, presumably on the strength of First World War experience with the Royal Flying Corps. His early work there resulted in the stabilised automatic bomb sight that was used by Bomber Command to sink the Tirpitz.

Transferred to the petroleum warfare department, Hartley was then approached by Bomber Command which was suffering appalling losses among aircraft returning to fog-bound British airstrips. The result was Fido – Fog Investigation Dispersal Operations – which cleared an area 1,000 yards long by 100 feet high to help more than 2,500 safe landings.

At the same time that Hartley was working on Fido, he was also developing Pluto – Pipe Line Under The Ocean. The aim in this innovation was to lay surreptitiously pipelines which would be able to feed petrol from Britain to mainland Europe, for use, eventually, by the Allied forces of liberation. Hartley remembered from his days in

PLUTO – Pipe Line Under The Ocean – was a brilliantly simple concept for delivering fuel to the continent by small-bore pipelines surreptitiously laid under the Channel from small ships

Iran that small diameter pipes had been used successfully to transfer petrol under high pressures for long distances. He proposed using conventional submarine electric cable for the task, omitting the copper. The idea met with both technical and military scepticism, but Hartley was proved right: in all, 23 Pluto pipelines were laid beneath the Channel, and with the later land-lines on the same principle more than a million gallons of petrol a day was transferred to the Allied armies as they advanced into Germany.

But the undoubted highlight of the first year of peace was the inaugural James Clayton lecture. Clayton had been the chief engineer of Courtaulds and a member of the Institution. In his will, details of which were announced in September 1945, he left the Institution £130,000 to be used "for the encouragement of modern engineering science". It was an extraordinarily large sum, equivalent to perhaps £2 million today. Previous bequests had only rarely reached into four figures, and much time at Institution council meetings was taken up with deciding the disposition of sums as small as £5 which had been donated as "prize funds" many years earlier. The "Engineering Applied to Agriculture Fund", for instance, consisted of £301-worth of Metropolitan Water Board B stock and generated an annual income of about £9.

James Clayton, chief engineer at Courtaulds, whose huge bequest is commemorated by the IMechE's premier lecture

The Clayton bequest stipulated that not less than a quarter of the income should be used for an annual prize; the council decided that some of the rest should go towards a series of lectures. Acting with commendable speed, it engaged Air Commodore Frank Whittle, the inventor of the jet engine, to deliver the first lecture on October 5, 1945.

Wartime restrictions on material which might be of use to the enemy had meant that little information had been published on the

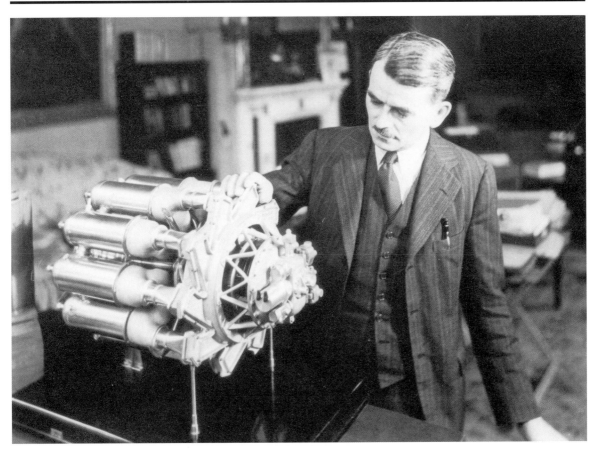

Frank Whittle and the jet engine: Whittle's IMechE James Clayton lecture in 1945 was so popular it had to be repeated in London and around the country

jet engine until very late in the war. Earlier, technical journals had made coy references to "work that cannot be discussed till the war is over". But in January 1944, a full 11 years after development work had started and seven years after the first engine had run successfully, a joint British and American statement, issued in Washington, announced that "jet-propelled fighter aircraft have successfully passed experimental tests and will soon be in production". The first successful flight, the statement revealed, had been in May 1941. The statement paid tribute to the "genius and energy" of Group Captain Whittle.

Accompanying Whittle's James Clayton lecture were photographs of the first jet engines. The basic view of the engine was "edited" to remove sensitive detail

Early Whittle engine under test at Lutterworth

Even after this announcement, full technical details of Whittle's engine were not released – though it later transpired that the Germans were far from ignorant about the jet propulsion work and had, by the end of the war, jet aircraft of their own. There was, therefore, considerable anticipation among engineers and among the wider public about the James Clayton lecture, and the fact that Whittle was to give his first public exposition of his invention at the IMechE was seen as a big coup for the Institution.

Whittle did not disappoint. In a remarkably open talk, he described the history of the jet from his first patent in 1930, through the formation of Power Jets in 1936 and the turbine blade and burner failures of the early war years, to the Gloster Meteor, the first British production jet-powered aircraft. He predicted that, in 10 to 15 years, all aircraft except light planes would be powered by gas turbines rather than piston engines.

The lecture was an enormous success. So great was the demand for tickets that it was repeated in London in January 1946 and then taken to 12 provincial centres, though not all the provincial talks were given by Whittle himself. Whittle was awarded the first £1,000 James Clayton Prize.

De Havilland Goblin jet engine from 1945 fitted to a Vampire aircraft

The success of the Whittle lecture contributed to a particularly buoyant period in the Institution's history. Across the second half of the 1940s, the Institution seemed more ready than in the past to debate and discuss new technologies and new industries, and aerospace was well to the fore, with discussion of not just new engines but also airframes, landing gear and instrumentation. This was the time when the wartime military aircraft developments were being scaled up to create the first generation of true civil airliners, and British engineers and British companies were well to the fore in the new technologies.

New industries were a big factor in the good health of the Institution, and contributed to an enormous growth in demand for membership. But some of the huge increase in membership in the years after the Second World War was the result of the absorption of the Institution of Automobile Engineers into the IMechE in 1947. The IAE staff, including its secretary Brian Robbins, who was later to be secretary of the IMechE, moved into offices on the top floor conveniently vacated that same year by the Treasury Solicitor's department, the last of the wartime evacuations from Whitehall to be accommodated in Birdcage Walk.

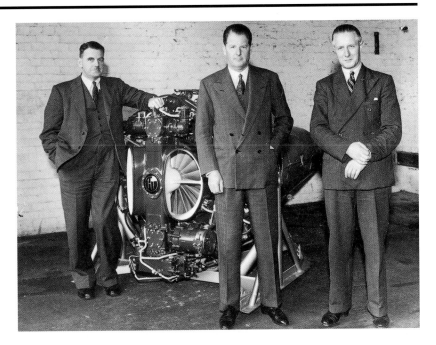

The Goblin team, Major Halford, Mr Brodie and Mr Moult, 1945

The amalgamation with the automobile engineers was not universally popular. There was a heated debate at the IMechE in November 1945, in which, on a show of hands, the motion for amalgamation was approved by only three votes, but a postal ballot of members voted by more than two to one in favour of the merger. A big worry was the creation of a separate, and largely autonomous, Automobile Division, the first division, as such, that the Institution had ever had, though some of the existing groups also had a lot of autonomy in terms of organising their own meetings and lectures.

A more important factor in the growth of the Institution, though, was the large number of former servicemen released on to the labour market at the end of the war and seeking qualifications. In 1946, the

Rolls-Royce's first gas turbine engine was the Welland, a direct development from Whittle's engine

Institution agreed to allow the setting up of short intensive courses to enable "suitable ex-servicemen to achieve Ordinary and Higher National Certificates in a reasonable time". The rules about workshop training for graduate members of the Institution were made more flexible so that those whose careers had been interrupted by war service would not be excluded.

Engineering was, of course, still overwhelmingly a male preserve as a profession, despite the large numbers of women who had been employed in the industry, largely in unskilled jobs, during the war. The Institution of Mechanical Engineers was no worse – but also no better – than other engineering bodies and companies in its male domination. The first paper to be presented by a woman had been back in 1918 – Miss Monkhouse's talk on "The Employment of Women in Munition Factories" had also been the first ever paper delivered at the Institution on labour and employment matters. The first woman member, Verena Winifred Holmes, was elected in 1924 and transferred to full membership during the Second World War, in 1944.

But among the hundreds of new members elected to membership or transferring between classes of membership in the postwar years, women still made up a very small percentage. Early in 1948, Miss M Poucher, a graduate member, delivered a paper to the Scottish section on "Fan Design". A year later Miss G A Sargent, a student member from Arundel, became visits secretary for the Graduate Section Southern Branch: she appears to have been the first woman to take up an elected post within the Institution.

Progress in this respect continued to be painfully slow. The annual report for 1961 reported that overall membership had passed through the 55,000 barrier during the year. Women members at that point numbered just 32, and of those only 10 were full corporate members.

In terms of absolute numbers, though, the postwar years were a time of enormous growth for the Institution. Total membership at the end of 1946 was 24,486, of whom 9,569 were graduate or student members; by 1952, the total stood at 39,570. Within 20 more years, the numbers virtually doubled again. Attendance at Institution events also increased dramatically in the postwar years, a phenomenon shared with other activities such as the arts and sport. Peacetime brought with it a huge growth in participation in a wide variety of activities.

Amid the growth and the progress, though, there was also introspection. The centenary of the founding of the Institution was marked in June 1947 with a week-long celebration. This included a thanksgiving service at Westminster Abbey attended by 1,100 members and their families, with the sermon given by the legendary "Red Dean" of Canterbury, Dr Hewlett Johnson, who had trained as an engineer before his ordination.

The celebrations provided evidence that engineering was regarded as highly important in the corridors of power. There was a huge centenary meeting in the Great Hall of the Institution of Civil Engineers, which was addressed by the Lord President of the Council, Herbert Morrison. The government held a reception for the Institution and its members at Lancaster House, and there was a banquet at the Guildhall in London at which Prime Minister Clement Attlee and the Lord Mayor spoke. Those not invited to the banquet could buy tickets for a dance and buffet at the Connaught Rooms.

Centenary celebration: IMechE president Lord Dudley Gordon unveils a plaque on the Curzon Street Goods Offices of the British Railways London Midland Region on February 26, 1948. The building, 101 years earlier, had been the Queen's Hotel, venue for the founding of the Institution

Morrison, effectively the deputy Prime Minister and in charge of the government's overall science policy, used his speech to the centenary gathering to announce that the IMechE secretary, Dr Henry Guy, was to chair a new Mechanical Engineering Research Board. This would identify post-war needs for basic research in mechanical engineering. Dr Guy had already been involved with the Advisory Council of Scientific and Industrial Research, to which the new board would report.

Another part of the centenary celebrations was a retrospective exhibition entitled "A Century of Mechanical Engineering" at the Science Museum which was opened by the education minister George Tomlinson and visited by 15,000 people a week for two

Mechanisation's influence in high places: Winston Churchill (seated), his successor Anthony Eden and his son-in-law Christopher Soames watch a Fergie tractor at work at Chartwell

Professor Andrew Robertson, IMechE president immediately after the war

months. In addition, 36 centenary lectures, covering all aspects of engineering and delivered by some of the top names in the profession from around the world, were held and 32 engineering companies and other organisations in the London region threw their doors open to IMechE visitors.

Celebration of any sort was not, in fact, particularly easy in the immediate post-war years. The restrictions of austerity limited the numbers at the 1946 annual dinner, and led to the outright cancellation of the 1947 dinner. The Institution's publications had been in some disarray since the start of the war, as paper shortages limited the publication of the proceedings; the shortages got worse, not better, in the post-war period, and the IMechE's intention to resume normal publication quickly had to be deferred. There were financial worries too: in 1948, there was the first instance of what, a quarter of a century later, would be a depressingly regular event, the special meeting called to approve an increase in subscription levels.

So the revival of enthusiasm which accompanied the end of the war and the growth in the Institution's membership and influence were tempered by the general mood of austerity. Professor Andrew Robertson of Bristol University, delivering the president's report to the 1946 annual meeting, remarked rather dourly that the difference between wartime and peacetime was "only that the removal of blackout regulations has made it easier and more attractive for members to attend meetings".

Austerity extended to the engineering industry which employed large and increasing numbers of Institution members. The Second World War had finally banished the lingering depression of the 1930s and wartime deprivations had created a pent-up demand for manufactured goods which should have more than filled industry's capacity. New technologies were available for exploitation; existing technologies were ready for much wider use. But peace brought new problems.

Other European economies devastated by the war were helped in the process of rebuilding by American aid under the Marshall plan: this outside help was not available to Britain. Relentless problems in

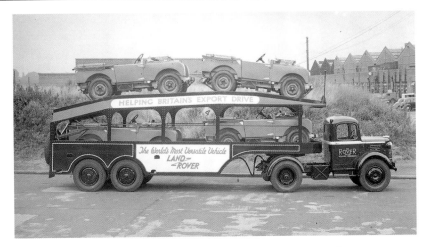

Export or die: the Land Rover, part of the post-war export drive

the wider UK economy across the late 1940s were met by a suppression of domestic demand through rationing, a continuous export drive, and the reorganisation of major industries which supplied raw materials and energy or which provided services such as transport and healthcare. Largely because of the prevailing politics of the time, the reorganisation was mainly by nationalisation, though in many cases some form of industrial restructuring would have been necessary regardless of the party in government.

The effect of all of these industrial and economic constraints was to divert attention and energies in industry away from innovation and technology. The burst of innovation that had characterised the Second World War and which had been in part responsible for securing victory was not sustained into peacetime as more mundane economic problems crowded in.

At this stage, the IMechE was only just beginning a long process of transition into a body which would, by the 1980s, put the role of the engineering discipline within industry at the top of its agenda and which would take for itself a leading role in encouraging innovation in industry. The postwar IMechE, at least at national level, was very much an Institution of individual members owing allegiance, though to the discipline and the notion of a wider industrial role was not

Some British cars had long-standing export markets: MG sportscars consistently sold well in the United States

clear. The war had given the IMechE an influential voice in national programmes of research and development; less certain was its exact relationship to industry, except as a body composed of individuals who in many cases worked in important positions in industry.

Lists of papers delivered or lectures given at IMechE meetings in the 1940s and 1950s frequently included "industrially relevant" topics, and there were plenty of contributions on management and employment issues, as well as descriptions of technical developments. But there was also an engaging haphazardness about programmes at national and branch level: almost anything could be covered, which reflected the broad range of activities encompassed by mechanical engineering, though some events patently owed as much to their social content than to their technical content. One of the highlights of the IMechE calendar was still the annual summer meeting, as ever held in a university or industrial town, at which social events dominated and the technical papers presented were frequently retrospective rather than forward-looking.

Within this atmosphere, of course, individuality, sometimes eccentricity, was encouraged and flourished. The postwar years saw a big increase in the number of papers presented on subjects such as aerospace and gas turbines, but there were also papers on cinematogra-

Mechanical excursion: members of the IMechE inspect a dragline at Stewarts & Lloyds in 1952

Imperial sunset: in the 1950s, but not for much longer, engineering bodies in parts of the former Empire still took their cue from British institutions. This is the first annual meeting of the Indian section of the Institution of Locomotive Engineers in 1956

O V S Bulleid, the last of the great steam locomotive engineers to be IMechE president

Merchant Navy class 4-6-2 nearing completion at Eastleigh Works in November 1944. O V S Bulleid, chief mechanical engineer of the Southern Railway, had been deputy to Sir Nigel Gresley in the great pre-war days of streamlined locos on the London & North Eastern Railway. The locomotive number, 21C11, was a further Bulleid innovation, denoting wheel arrangements: under British Railways, this was No 35011, General Steam Navigation

phy and quite a few branches, not just in the graduate sections composed of younger members, seemed to find visits to breweries of value. Technology was covered, of course – indeed, a prime role of the Institution always had been the dissemination of technical information – but the Institution was also a social and educational establishment, and, in some respects, a kind of gentlemen's club.

Some of the subjects covered in meetings and lectures owed more to the past than the future of engineering, and distinguished engineers seemed able always to command a platform for their views, even if unorthodox, at the IMechE.

The postwar years saw the last of the great steam locomotive engineers who form a continuous thread of membership of the Institution from George Stephenson in 1847. The president of 1946/47, O V S Bulleid, had been Sir Nigel Gresley's assistant during the days of the big streamlined locomotives, including the record-breaking Mallard, on the London and North Eastern Railway. Bulleid himself, as chief mechanical engineer of the Southern Railway, built the successful Merchant Navy and West Country classes of Pacific locos.

Bulleid used his presidential address to advance the contentious proposition that the steam locomotive still had a lot of potential for further development and could be made to rival other more modern forms of traction, even electrification, the long-term favourite of most engineers, for efficiency. Many of the steam engine's difficulties, he said, were to do with the quality of coal, and he supported the conversion of steam engines to oil burning. He also advocated a radical redesign of the locomotive to achieve the new efficiencies. It was not to be.

But though some papers, like Bulleid's, betrayed an occasional hankering for past engineering glories or pursued technologies which, not just in hindsight, were of dubious importance, many other contri-

Bulleid's Leader class locomotive 36001: the only one of the class completed, it embodied many of Bulleid's ideas about the perfectability of the steam locomotive, but perhaps had just too many innovations. It is seen here in Seaford Terminal station in August 1949

butions in the immediate post-war years showed that the Institution was right at the leading edge of technology and innovation.

In the middle of the naturally retrospectively-inclined centenary celebrations of 1947, the third of the James Clayton lectures was delivered by Professor J D Cockcroft – later Sir John Cockcroft – of the newly-created Atomic Energy Research Establishment at Harwell. The subject was "Nuclear Energy for Heat and Power Production". It was fewer than five years since Fermi had built the first nuclear reactor in the squash court of the University of Chicago; fewer than two since the detonation of the first nuclear weapons, to the development of which Cockcroft had contributed from his post as wartime director of the atomic energy division of the Canadian National Research Agency. IMechE members clearly were being kept right up to date with developments.

To modern eyes, Cockcroft's lecture comes across as a model of openness. The construction costs for a 75 Megawatt nuclear power station would be about 2.5 times those for a coal-fired station, he said; operating costs might be about 25 per cent higher, though there was uncertainty about this "owing to our lack of knowledge of the amount of fuel recycling required".

Already, the problem of nuclear wastes had been identified: "We shall have to solve the problem of the disposal of radio-active fission products which would be produced as a by-product in very large intensities. These products can be a serious danger to health if they are dispersed in concentrated form."

Cockcroft concluded that researchers would need at least a decade to investigate the technical problems. "All we can say now is that it is worth a determined effort and that effort we will make."

Rather less than a decade on, IMechE council member and future president Sir Christopher Hinton, later Lord Hinton of Bankside, delivered another James Clayton lecture on "Nuclear Reactors and Power Production". At this stage – 1954 – Britain had built not only the experimental atomic pile BEPO (British Experimental Pile) at Harwell, but also the first production-scale reactor at Windscale in Cumberland.

Secrecy about nuclear matters had increased in the intervening years since Cockcroft's lecture, but Hinton neatly sidestepped the problem. "A fairly full description has been given of BEPO," he said, "because most details of its construction have been declassified. The Windscale piles cannot be similarly fully described, but they can be understood by remembering that in principle they are similar to BEPO."

The start of a new industry: Britain's first nuclear power station at Calder Hall under construction in the early 1950s

Like Cockcroft seven years earlier, Hinton made predictions about the likely development and costs of nuclear power generation. The cost of electricity from a prototype reactor, he said, was "in the neigh-bourhood of a penny a unit", which was "about 50 per cent higher than the average cost of generation from coal-fired power stations". He thought, though, that the price would be bound to come down in larger nuclear stations and that the price of coal was likely to rise: "It would not be surprising if nuclear power became fully competitive within another ten years."

He concluded: "It would seem wise to go ahead and build a few large installations which would give us the necessary operating and engineering experience for future plants capable of generation at lower cost."

Lord Hinton of Bankside's lifelong commitment to the IMechE, which included leaving all of his papers to the Institution on his death

New industrial landscapes: the nuclear power industry brought new shapes to the sky-line, as here at Windscale

in 1983, began with his early mechanical engineering career at Swindon on the Great Western Railway. In 1923 he was selected from 110 applications by the Institution to receive the William H Allen grant which had been set up to fund students of limited means through the engineering honours course at Cambridge. Hinton's later engineering career brought him prominence in munitions, electricity generation and nuclear power: he is a prime example of the breadth of mechanical engineering.

Coincidentally, one of Hinton's colleagues on IMechE councils in the 1950s and 1960s was Norman Allen, president in 1965 and head of the Bedford-based engineering company W H Allen, founded by Hinton's original benefactor. Norman Allen was the fourth generation of his family to be a member and made it clear in his presidential address that the IMechE was as much part of his heritage as the family firm.

Many IMechE members were also members of other institutions which dealt with more sectional engineering interests. There was, for instance, a long history of dual membership of the IMechE and the Institution of Locomotive Engineers before the two merged in the late 1960s, and several of the ILE presidents became president of the IMechE too.

Where the biggest engineering institutions – the Mechanicals, the Civils and the Electricals – differed from the smaller more sectional engineering institutions at this stage was in the breadth of topics covered, and this included the provision of instruction in areas where there was as yet little in the way of formal qualification, such as organisational theory, management and business studies. The IMechE had for many years required all members, in the examinations which tested their competency, to complete a paper on the economics of engineering. While some university engineering courses included tuition on management and business matters from prewar days, others remained resistant until relatively recently.

As they had done since the changes at the end of the First World War, management issues formed an important strand in the subject matter for IMechE meetings both at national and local level in the postwar years. One of the first papers delivered at the Institution in the month after the end of the war in Europe in 1945 was by two managers from the Ministry of Supply's Royal Filling Factories, describing a time-and-motion study scheme for maintenance work involving an incentive bonus scheme.

Organisational theory and management practice bulked large in the formal papers given in the period before conferences became a more normal method of disseminating information to members. Some of the early lectures obviously reflected a sectional interest of members, or the personal views of the speaker. In 1948, the automobile division heard a talk on "The Organisation of Large-Scale Vehicle Overhaul and Repair", which can scarcely have been a subject of universal concern for mechanical engineers; a year later, the Institution as a whole was treated to the views of H E Merritt in the delightfully titled "Simplification Creates New Problems for Top Management".

Over time, the lack of formal management training in the wider education system for engineers would become a cause for frequent complaint in Institution meetings, and the complaints were often matched by the converse grumble about the apparent lack of engineering knowledge among industry managements. Sometimes the two threads of dissatisfaction came together in more general discussion about whether engineers should have to leave engineering behind to progress their careers in management.

The emphasis on management topics within the IMechE in this period was natural in any case because so many new management concepts were introduced through engineering operations in manufacturing companies. Engineering departments were early candidates for quantifying productivity because output could be measured more easily than in companies in service sectors. Ideas such as quality control were similarly introduced early into engineering companies, and the IMechE was running conferences and seminars on product quality and reliability from the early 1960s onwards. Concepts such as value engineering were brought in from the United States and modern methods of purchasing supply chain management filtered out from the big automotive and aerospace companies into general engineering practice.

The breadth of discussion in the years up to 1960 was part of the tradition of the Institution of Mechanical Engineers as a "broad church", which dates back to the earliest days. In the 15 years after the Second World War, however, it became clear that the Institution had to align itself with demands from both government and industry for more formal qualifications for engineers. Industry was calling out for more engineers of all kinds; successive governments were committed to providing, through the education system, a big increase in the numbers of qualified engineers. The engineering Institutions, the IMechE included, were invited to contribute in terms of setting standards and accrediting courses, and the strong inference was that they should bring their membership criteria into line with the wider education system.

The education of engineers has always been one of the central roles of the Institution of Mechanical Engineers. The broad church tradition, however, meant that the Institution prided itself on the variety of ways in which membership could be achieved and on the way

Alfred Roebuck, IMechE president from the practical tradition of mechanical engineering, who questioned the drift towards an academically qualified profession

in which engineers with few formal qualifications could still be members of the Institution. The tradition of members with little formal learning but immense practical knowledge dated back to the founding of the Institution and to its founding father, George Stephenson.

The moves in the 1950s towards greater formalisation of engineering education started a debate within the Institution which has echoes to this day. The traditionalist viewpoint was put by Alfred Roebuck, a Sheffield steelmaker who had risen to the top of the IMechE with no formal educational qualifications but a vast experience of engineering. His presidential address in 1952 stressed the links between the engineering profession and the equally long tradition of engineering craftsmanship.

He said: "One is impressed by the high degree of craftsmanship attained by famous engineers such as Watt, Stephenson, Maudslay and Nasmyth and, notwithstanding their eminence as engineers, they knew nothing of National Certificates, Endorsements, and PhDs. It is possible that today we have gone too far in the other direction, since it appears that scientific and technical qualifications are often quoted as a measure of the worth and value of an engineer without due regard being given to his knowledge and skill in the actual practice of engineering."

Behind this was a long-running and never-to-be-entirely-resolved debate about the nature of mechanical engineering. Was it a science? Was it a craft? At different stages across the 150 years of the Institution of Mechanical Engineers, the balance between the theoretical and the practical elements of engineering has been difficult to define and has shifted constantly. At one point in the 1950s, the craft tradition that Roebuck exemplified gained the upper hand to the extent that the use of the phrase "engineering science" was openly questioned.

The expansion of engineering education – and the obvious desirability for the Institution to contribute to the setting of the standards and to align its own qualifications with the qualifications given by the wider education system – brought the classroom versus workshop debate into the open. It also put the "learned society" function of the Institution on a more formal basis.

From its earliest days, the Institution had a direct role in disseminating information about innovation and technology. Until well after the Second World War, the methods of dissemination were pretty much those of a Victorian Institution. The IMechE was organised on both national and local levels, and in both there were programmes of events: lectures, discussion papers, regular meetings, visits. The sectional interest groups that were formed in the 1930s dealt with specific parts of mechanical engineering, but again, their methods of disseminating information were the essentially Victorian ones of formal meetings and lectures. The more important papers from all of these Institution meetings were collected together each year in the Proceedings. There were other publications too, in pamphlet, book or, increasingly, journal form and the headquarters of the Institution in Westminster housed the library and the archives, which members could use for their researches or for keeping themselves up to date with technologies.

In innovation and technology, the Institution had traditionally offered other services. One of these was the technology adviser service, in which the Institution retained engineering experts who were available, at short notice, to help members with their own inventions. The service was discontinued after 1945, when the president reported that no members had availed themselves of it in the previous year. A different role dating back to 1879 was the support of research projects at universities and polytechnics with cash grants; this resumed after the war, but had dwindled to a very small operation by the mid 1950s.

The past 50 years, though, have seen more changes in the methods by which the function of disseminating information about technology and innovation is achieved than there were in all of the first 100 years of the Institution's existence. The changes have involved the development of a huge business in conferences and seminars, a wholly new publishing business, and – in recent years – the use of new technologies in information provision.

The conference business was to a degree a logical extension of the earlier lectures and meetings. A few conferences – sometimes called symposiums or seminars – had been held before the war, and a few, often in collaboration with other engineering institutions, were even run during the war.

But the concept of a day, or longer, devoted to a series of papers on one subject was still relatively untried until the 1950s. For success, it relied on the co-operation of employers, who had to be persuaded that releasing engineers to attend was an investment that would be repaid by the knowledge the engineers would pick up. The art of running conferences therefore was to select subjects that would appeal to engineers as individuals and to their employers too: technology-oriented topics were likely winners, but even more certain were conferences on broad technical issues in innovative design and manufacture.

The increasing conference programme of the Institution also fed into the new publishing business. With membership increasing rapidly and with the Institution becoming more active in areas such as education, the annual proceedings and ad hoc communication through the branch structure was no longer sufficient. The Institution of Automobile Engineers had already set up a monthly journal before it merged with the IMechE; the IMechE itself had set up a journal in 1914, and given it an expanded role in the 1930s, partly to relieve the pressure on space in the Proceedings, which was still regarded as the premier publication. But the journal was still largely a noticeboard, and with increasingly professional technical journals available elsewhere, was beginning to look old-fashioned.

It was a logical decision, then, in the early 1950s, for the Institution to go into publishing properly, with a monthly magazine which would become the main means of communication with members, and a book publishing venture which was able increasingly to work alongside the conference programmes.

One effect of these moves towards creating within the IMechE new "businesses" in conferences and publishing was to throw the emphasis of the Institution very much back on to innovation and technology. Both the conference programme and the revised journal – called Chartered Mechanical Engineer – thrived on technical information and innovation in design, manufacturing and process.

New business: as part of the IMechE's new sense of direction in the 1950s, the Journal, *which gave rather formal announcements of events, was replaced by a magazine,* Chartered Mechanical Engineer

Foreword by the President

THE NEW PUBLICATION

I N PLACING BEFORE MEMBERS this first issue of *The Chartered Mechanical Engineer* the Council are giving concrete expression to their recent decision to modify their publications policy. It is their sincere desire that this new publication, which will be issued monthly (excluding the months of July and August), shall prove informative and interesting to members. They believe that it will be the means of making available to mechanical engineers valuable information which does not fall strictly within the scope of the existing publications. At the same time they intend that the standard of this new publication shall be such as to enhance rather than merely to maintain the dignity and prestige of the Institution. The Council will be glad to consider any further suggestions for achieving their object.

The various publications of the Institution form an important means of realizing one of the main purposes for which the Institution is constituted, namely, to facilitate the exchange of information and ideas on mechanical engineering. It follows that the Council carry the responsibility for maintaining these publications at the highest possible standard consistent with the financial resources available.

The latest decisions of the Council will, they hope, be welcomed by members as continuing the progressive publi-

by the Council after very careful study and analysis, that there is a body of general readers who would prefer to have summaries of papers rather than the papers in full, provided that these summaries set out the main arguments advanced and the main conclusions reached in each paper. The Council have therefore decided to publish summaries of all papers accepted for publication in the PROCEEDINGS, these to be of sufficient length to give a clear indication of the content of the paper. The summaries will appear in *The Chartered Mechanical Engineer* so that members will be

It was a good direction for the IMechE to be taking in the 1950s and into the 1960s for there seemed to be a revival in the Victorian faith that technology was one of the main keys to progress. The revival of this view was altogether more sanguine than the devout faith in progress of a century earlier. The idea that technology could be a key to civilisation had to be seen alongside the knowledge that, only a few years earlier in the Second World War, technology's capacity to create the means of destruction had been starkly revealed. Some of the new enthusiasm for the concept of progress found in the mid to late 1950s was heavily overlaid with moral overtones: ideas like Atoms for Peace stressed the non-military uses of technologies derived from wartime.

The 20 years after the end of the Second World War did, though, mark a period of huge technological change. Beginning slowly from the austere years of economic straitjackets after the war, mechanisation moved into the home with the development of mass markets for labour-saving domestic appliances; motor car sales, which had stalled through the depression of the 1930s, bounced back, with UK production topping the one million mark for the first time in 1958 and car ownership becoming, by the mid 1960s, the norm instead of the preserve of the wealthy; new or unfamiliar words, such as computer, automation, robot and productivity were starting to be heard in industry.

With the lid off some of the technologies developed in war-time, the Institution heard an awful lot about aero-engines, jets and gas turbines in the early years after 1945. The new industry created by Whittle demanded innovations in design, in manufacture, and in metallurgy, and all were debated and discussed at the Institution. So were alternative uses for the technology.

British Railways' engineers, searching for the locomotive which could bridge the gap between the ageing population of trusted but less than efficient steam engines and the future unaffordable nirvana of a fully electrified rail network, encouraged commercial locomotive

Domestic comforts: Hoover's Mark I washing machine, with handy in-built mangle, took much of the sweat out of a regular household chore

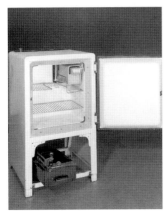

Affordable luxury: paraffin-fuelled refrigerator for use in prefabs and flats, made by Electrolux at Luton

builders to develop gas turbine engines. They worked, but not well enough to deflect BR from a policy of going for diesel traction once it was satisfied that its earlier objections over pollution and cost had been overcome.

There were similar projects to create a gas turbine powered car: one, with the registration plate JET 1, was built in the familiar body-work of the wedge-shaped 1950s Rover and generated an unfamiliar 200 bhp. Again, the ideas lingered for a while, but eventually dropped from view. But as late as 1974, General Motors' group vice-president E M Estes, delivering a James Clayton lecture on "Alternative Power Plants for Automotive Purposes", rated gas turbines third after conventional gasoline engines and diesel engines – and ahead of battery and fuel cells – as a potential future motive power source for cars.

Other motoring innovations from the 1940s and 1950s stood the test of time better. The first tubeless tyres for cars were announced in 1953 by Dunlop, the company which had fitted disc brakes to Jaguar racing cars the year before; automatic gearboxes and overdrives,

Cult car: the millionth Morris Minor comes off the Cowley production line that started in 1948

used on top-of-the-range models before the war, started to be produced for mass market cars; new materials such as glassfibre and aluminium, started to be used for bodywork and for components.

The motor industry was one of the first worldwide manufacturing industries to be dominated by companies which were organised on a worldwide basis: technologies developed across the Atlantic or in mainland Europe quickly found their way on to British-built cars or into British factories, and the Institution was an important part of the process of spreading the word. Four-wheel drive, power steering, hydrolastic suspension and the automotive diesel engine were developments which were taken up first by overseas motor manufacturers, but which automotive engineers in Britain were quickly made aware of.

The motor industry was also a continuing source of manufacturing

By the late 1950s, railway doubts over diesel traction had been overcome: a Warship class diesel hydraulic loco heads towards Paddington

*The pioneer Deltic diesel-
electric locomotive*

innovation as well as new technologies. Earlier in the 20th century, Ford had brought the concept of mass production to Britain, and in the postwar years the techniques of flow-line manufacture and assembly were applied to many other manufactured goods, notably domestic appliances. Car manufacturers had also pioneered the automation of processes such as welding in the inter-war years, and their work practices had had profound effects on industrial technologies such as mechanical handling and machine tools.

*The end of an era: Britain's last
new-built mainline steam locomotive, the British Railways
2-10-0 no 92220 Evening Star,
named at Swindon on March
18, 1960*

*The opening of the first motor-
ways in the late 1950s brought
the first real alternatives to
railways for inter-city travel*

For much of the post-war period, though, automation meant
process control systems rather than production technology: it was
only towards the end of the 1950s that the term, American in origin
and often spelled with "inverted commas" around it, was discussed in
connection with machine tools in any depth.

Process control also formed a long-running concern of the
Institution of Mechanical Engineers, dating from the immediate post-
war period, though the automatic control group was only set up in
1961. The IMechE, though, had been talking about automatic process
control for many years before that; the analogue computing pioneer
Arthur Porter from the University of Manchester had given a paper on
"Basic Principles of Automatic Control Systems" in 1948, and the following year had seen a paper on applying frequency response analysis to automatic process control.

The first numerically controlled machine tool had been developed
in 1952 at the Massachusetts Institute of Technology under a US Air
Force research programme: for much of the next few years, there
seemed to be an assumption in Britain at least that the idea would be
of limited application – primarily in the aerospace industry and principally on milling and turning machinery. The IMechE was keen to
keep its members abreast of the developing technology, though the
lead was taken by a different Institution, the production engineers,
which organised conferences on aspects of production automation
from as early as 1955.

Pinpointing the first use within the portals of Birdcage Walk of the
word "computer" – or, more likely, "computor", the standard spelling
until the mid 1950s – is like the proverbial search for needles in
haystacks. Certainly by the time D T N (Theo) Williamson delivered a
seminal paper on "Automatic Control of Machine Tools" to the
IMechE in January 1958, the word appeared to have passed into common usage.

Williamson, at that time head of the machine tool control department at Ferranti in Edinburgh, spoke about the number of calculations which had to be made to produce a punched tape or punch

Rover with a difference: one of the cars converted to a gas turbine engine in the postwar enthusiasm for fitting the jet engine to other forms of transport

cards to control machine tool movements. "There is," he said, "an irreducible minimum of calculations which must be made. For other than simple work such as template manufacture, which consists of two-dimensional interpolation, these data can be quite considerable and their handling can be responsible for the major operating cost of a numerical machine-control system, especially where the number of components to be made is small." American experience, he said, indicated that hand computation took 20 times as long as a computer would take. "On average," he concluded, "the use of a computer will allow the number of planners necessary to operate a system to be divided by 10."

The significance of this statement is not so much in the technology as in the consequences of its use: the IMechE was hearing, if it cared to listen, a very early indication that computers would radically affect working lives. Over the coming years, the Institution would distinguish itself by the degree to which it was willing to debate the social consequences of new technology, as well as the technicalities. Of some significance to manufacturing engineers as well was Williamson's view that the savings in manpower were particularly important in small-batch manufacture: this view led him on, in the 1960s, to develop the world's first flexible manufacturing system for Molins, bringing automation to small batch manufacture.

At this stage, of course, digital computers – as distinct from calculating machines descended from Babbage's mechanical calculating engines of 100 years earlier – were complex arrangements of valves housed in enormous cabinets and requiring delicate treatment. The Americans always claimed to have developed the first true computers. IBM built an experimental machine in 1944 with 750,000 components and 500 miles of wiring and the Electronic Numerical Integrator and Computer ENIAC was a "working" computer built by the University of Pennsylvania during the war – though not completed until 1946; it was used to calculate trajectories of missiles fired from guns. ENIAC had 18,000 valves and took up 1,700 square yards of floor space.

In fact, it is likely that the British had the first working computer: the Colossus machine built to crack the German Enigma codes at Bletchley Park was operational in 1944, but because of official

Here's looking at you, kid. The 1939 De Havilland Flamingo, the epitome of pre-war, pre-jet engine Bogart and Bergman flying

secrecy, little was known about this for the next 30 years. Britain had a second claim to fame in the computing pantheon: it had been work at the University of Manchester which had taken on the problem of data storage and which had led to the world's first commercially available computer, the Ferranti Mark 1 Star.

By the end of the 1950s, the first stages of the process of miniaturisation which would revolutionise computing were starting to take place: transistors were replacing the original valves. Nevertheless, at the time the Institution of Mechanical Engineers held its first meeting specifically about computers in production, on the first day of spring in 1960, there were probably no more than 200 computers of all kinds in the whole country.

The meeting was not just the first on computing: it was also the first in "informal discussion" format, in which topics were introduced by several speakers and then opened up for general debate. Sir Christopher Hinton, by now vice-president of the Institution, explained that the new format was because the topic for discussion was not new work, but was "not broadly known to engineers". The informality seemed to go down well: the meeting hall at Birdcage Walk was filled to capacity with more than 300 members.

The discussion was sensible, serious, and surprisingly free of the kind of misconceptions liable to make participants cringe with embarrassment as they read the report in later years. The speakers, mostly from the newly-developing computer industry, with a sprinkling of people already using computers in factories for office routines or production control, all talked in plain terms, avoiding jargon. The unfamiliarity of most engineers with computers meant that some basic definitions had to be made.

What a difference a decade makes. The 1949 De Havilland Comet prototype: the world's first jet airliner and the pioneer in an aviation revolution

One speaker, P J MacLeod of ICT (International Computers and Tabulators), said he wanted to dispel the "black box" image of computers. "Black box usually means magic," he said, "and that usually means not only secret delights but also some secret failures. You get the idea that the computer is a sort of 'Big Brother' who is watching you and would, if your efficiency drops a tenth of 1 per cent, tell someone, and what's more tell someone behind your back." This might be a common experience in industry, he added, but you didn't need to buy a machine to do it.

"Computers are simply machine tools for calculation," he said. "Like most machine tools, they are very simple in conception and deliriously complicated in design."

The discussion came to no firm conclusions. A W J Chisholm, the chairman of the Industrial Administration and Engineering Production Group which had organised the meeting, wound up the debate by declaring that "mechanical engineers are clearly committed to greater and more universal use of the computer in the future".

Indeed, they probably were. And after the discussion the IMechE was also committed, by the success of the meeting, to the idea of holding more events of a less formal nature. In fact, the successful discussion meeting came in the middle of a prolonged period of quiet but determined upheaval for the Institution, out of which emerged the modern IMechE.

6

Highs and Lows

1960 TO 1980

T here is no precise date when the Institution of Mechanical Engineers began the process which would see it change into an industrially-focused business-like body. There had been structural changes at the end of the First World War, with the setting-up of the regional branches; the first special interest groups had been created in the mid 1930s; the absorption of the Institution of Automobile Engineers in 1947 had led to the formation of the first Institution division.

Nor is it an easy task to put a date on when the process of transformation was complete – inasmuch as it ever can be complete. A further batch of substantial changes was brought into effect in the early 1980s, and this severed some links with the past and put in place a new forward-looking technical structure and business organisation.

But there is no doubt that the early 1960s were years of transition and a period when the IMechE moved up a gear in terms of being a proactive rather than a reactive body and in terms of the national role that it took.

Some of the factors in this transition were to do with necessity rather than any conscious decision: membership had more than doubled in the 15 years following the Second World War, and the sheer volume of the work meant the IMechE had to put matters such as its publications and its programme of meetings on a more business-like footing. With double the membership, and more than double the number of meetings and conferences, the Institution was in any case turning into quite a sizable business, and communication with the members demanded a more professional organisation.

But there were also external influences. Governments of both main political parties in the post-war years had identified technical education and the numbers of qualified engineers in industry as issues central to economic performance and national competitiveness.

Industry itself was also becoming more vocal in identifying its needs. New competition for British manufacturing companies was coming from mainland Europe, North America, the Soviet bloc and beyond, and the traditional markets in the former Empire were loosening their ties with Britain and exercising their right to buy goods from other sources. Many British companies were discovering that they were being overtaken in terms of size, industrial muscle and technology by competitors from other industrialised countries, many of which appeared both nationally and corporately to give greater priority to engineering.

From within and from outside the IMechE, there developed across the 1950s a strong feeling that the engineering profession and the

Sir George Nelson, the chairman of English Electric who was later created Lord Nelson of Stafford, was successively president of two of the great engineering institutions, and was a powerful voice for engineering in the postwar debates about education

institutions which represented it needed to be more proactive, more organised and more aligned with the needs of industry. Sir George Nelson, the chairman of English Electric who was successively president of the Institutions of Electrical Engineers and Mechanical Engineers in 1956-58, articulated some of the new mood in his IMechE presidential address in October 1957.

Because of rapid advances in technology, Sir George said, engineering and science had become remote from everyday life. "As recently as 1890, for example, the Prime Minister Lord Salisbury had a laboratory at Hatfield House where he could dabble in his spare time. On the other hand in the last 100 years a gulf has grown between technology and the humanities which I can best illustrate by mentioning that many eminent men regularly raise after-dinner laughs by parading their ignorance of the elementary facts of science, whereas people regard with scorn rather than amusement the technologist who is completely ignorant of the arts."

Sir George urged engineers to use universities, schools and the media to spread the message that science and engineering were "essential to the maintenance and advance of our standard of civilisation in the future". The Soviet Union, he noted, attracted its brightest young people into science and engineering by paying its technologists better than other professions. The numbers of professional engineers in Britain had doubled since 1945, but a 1956 government report on scientific and engineering manpower indicated that there would be a shortage in industry of trained engineers of more than 30 per cent in industry by 1959. The new target was to double numbers again by 1970: "Personally, I would like to see it reached much earlier," he said.

Engineering in control: the control room of an early nuclear submarine indicates progress in mechanisation and automation

Sir George's address was one of the first openly "political" speeches in the Institution's history. It brought into the open themes such as the continuing need for greater numbers of engineers and the requirement for increased respect and status to attract the brightest and the best into the engineering profession.

There was, within the address, implicit criticism of the British attitude to engineers and engineering and of the prevailing culture which appeared to undervalue industry, production and manufactured products. But there was also implicit criticism of engineers and their institutions for allowing these attitudes to arise and for not doing enough to counter them. And there was a tacit admission that past debates within the Institution about the balance between craft and science in the profession and about the merits of graduate and non-graduate routes to engineering qualification had diverted attention away from the real issues.

These themes had first been raised in the days of engineering self-doubt in the First World War, but had largely subsided since. This time, they would reverberate across the politics of the engineering profession for the next four decades.

Almost as an aside, Sir George also, in passing, identified a hitherto underrated reason for increasing the numbers of scientists and engineers which would assume great importance in the rest of the 20th century: the depletion of the world's natural resources. Environmental concern was a fairly new issue for the immediate post-war years, and in the 1950s discussion was mostly confined to pollution control measures such as the Clean Air Acts which finally banished the city smogs.

Modern power stations such as Ratcliffe-on-Soar had to meet more exacting standards on pollution control and on using finite natural resources

But the rate at which mankind was using up the earth's resources would, eventually, lead to big businesses in recycling and process optimisation and, though Sir George appeared not to see the pressures that all industries and engineers would come under not to use scarce materials, he did forecast correctly that there would be a huge demand for engineering skills in devising replacement materials.

It was maybe not as a direct result of Sir George's address, but the IMechE began at the end of the 1950s to move forward in terms of projecting a more positive and dynamic image, and it started the first steps in taking the lead on engineering and industry issues.

In the early days of the new direction, it took the initiative in representing the engineering viewpoint on issues of national importance, such as the massive increase in university education which followed the Robbins report in the 1960s. This was part of a massive expansion of higher education in Britain one of the aims of which was specifically to help meet industry's need for engineers.

The expansion of higher education dating from the 1960s would lead, by 1997, to diplomas in completely new subjects such as utilities management

But it was not just in the public arena that the IMechE was taking a lead in raising the profile of engineering and aligning the profession's interests with those of industry. It took a long hard look at its own organisation and structures, and at those of the engineering profession as a whole, and began a process of change which continues to this day.

The stated aim of the reorganisation of the IMechE's group structure which came into effect in 1961 was to put the emphasis on industrial relevance rather than on engineering as a branch of scientific knowledge. The eight existing groups at the end of the 1950s dealt with applied mechanics; education and training; hydraulic plant and machinery; industrial administration and engineering production; internal combustion engines; lubrication and wear; steam plant; and nuclear energy. At the end of 1960, they were joined by five new groups covering automatic control; railway engineering; manipulative and mechanical handling machinery; thermodynamics and fluid mechanics; and process engineering, refrigeration, ventilation and vacuum plant.

The groups – and the automobile division, which, as the successor to the former Institution of Automobile Engineers, had always retained a lot of autonomy – were responsible for organising their own events, with the Institution as a whole, led by the council, vetting the ideas and co-ordinating the groups, and also taking charge of the big "set-piece" events. The restructuring of the groups had the immediate effect of increasing dramatically the numbers of conferences and symposia, with some groups deliberately aiming from the outset for at least one big conference a year.

With hindsight, the new group structure looks to have been scarcely a radical move. For the first time, however, the IMechE deployed its publicity machine, particularly the monthly magazine Chartered Mechanical Engineer, to advertise the new groups and to explain the rationale. The explanation was strangely complicated. The aim of the new structure, said CME, was to create two types of group: "horizontal" groups dealing with elements of mechanical engineering science which would appeal to all mechanical engineers, and "vertical" groups covering specific engineering sectors which were expected to be of more limited appeal and which would be added to as new disciplines emerged. In this explanation, automatic control was horizontal and process engineering vertical.

Fortunately, the members understood. In fact, the new group structure was an instant and sensational success. The original eight groups had had a total of 6,684 members – even if there were no Institution members who belonged to more than one group, this represented only 12 per cent of the total membership, indicating a huge degree of torpor.

But by the end of 1961, their first year of operation, the revised 13 groups had 41,430 members, an increase of more than six-fold. The introduction of the new groups revived the older groups too: the lubrication and wear group, for instance, suddenly increased in membership from 426 members to 2,810. The new thermodynamics and the automatic control groups instantly became the Institution's third and fourth largest groups, indicating that these were areas where members' interests really lay.

This restructuring was by no means the last time the IMechE would look at its groups and divisions, but the 1960 reorganisation set the tone in that it started a process of making industrial relevance and technology rather than pure engineering science the prime focus. Later changes reinforced this sense of direction.

There was a wider consequence for the Institution: the success of the restructuring in bringing large numbers of mechanical engineers into active participation was a huge boost to the collective self-confidence of both the Institution and the profession. It also set in place the perception of the Institution as a wide-ranging body, with a broad definition of mechanical engineering. The degree of autonomy given to the new groups to arrange large-scale events and to forge their own links with other organisations gave the IMechE almost a federal structure. It also emphasised the multidisciplinary nature of engineering and the universal importance of mechanical engineering.

The main thing, though, was that the members were keen on the new groups. But there were other activities that they increasingly demanded from a newly proactive Institution that the IMechE found more difficult to provide. In a time of increasing unionisation, professional engineers, either willingly or through the coercion of the closed shop, were seeking workplace representation.

Most members appreciated that the Institution's charitable status and Royal charter forbade it from acting as a trade union in any sense. But a debate rumbled on for the best part of 20 years, often involving other institutions and the umbrella body, the Council of

IMechE president Sir Kenneth Hague became the first chairman of the new Council of Engineering Institutions, formed to represent the whole profession

Engineering Institutions (CEI), as well, about whether the Institution should endorse union membership, or any specific trade unions, or whether it might not set up the machinery to allow it to act for members in some circumstances.

In the mid-1970s, when consultation with the recognised trade unions was made a statutory part of the process of nationalisation of the aerospace and shipbuilding industries, the IMechE did briefly step in to ensure that the many thousands of professional engineers who were employed in these industries but who were not members of any recognised unions had their views taken into account. But otherwise it retained the non-involvement required by charitable status and the Royal charter.

Even so, the union issue was one of the main debating points in the engineering institutions as a whole for many years, and the IMechE was involved in a series of policy statements which had to tread a careful political line. These at various stages included recommending that professional engineers should join a trade union and even went so far as to nominate appropriate unions for members to consider. And its proper refusal to take on a union role itself surfaced from time to time as a complaint in the letters pages of its magazines. As late as the Finniston inquiry into the engineering profession at the end of the 1970s, when the CEI held a series of regional meetings to help the Finniston committee members gauge "grass-roots" engineers' views, the lack of support for individual members in industrial disputes was the single strongest complaint against both the CEI and the institutions.

The CEI was itself a creation born of the new mood of self-confidence among engineers in the early 1960s – with some prompting from both government and industry. The IMechE took a leading role in the discussions with a dozen other engineering institutions which resulted in the setting up in October 1962 of the Engineering Institutions Joint Council, later renamed the Council of Engineering Institutions. The CEI's first chairman was Sir Kenneth Hague, a member of the IMechE, and the secretariat was originally at Birdcage Walk.

The Council of Engineering Institutions, formed in the early 1960s, brought out its literature under a "modern" logo

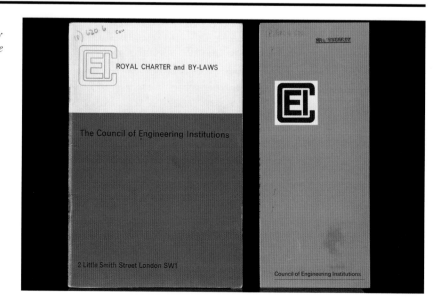

The CEI was the first organisation to span virtually the whole of the engineering profession. The presidents of the Big Three institutions – the Civils, the Mechanicals and the Electricals – had been members of a joint council for 30 years to 1952, but it had never had any powers or responsibilities, and it appears to have simply lapsed.

By contrast, the CEI was set up with the express aim of bringing some unity to the profession as a whole. It was intended to be permanent, and within three years it had its own offices and its own Royal Charter: more than that, it had the kudos of having as its first president Prince Philip, who had long taken an informed interest in engineering and technology.

The CEI was not without controversy, though, and its role was not universally appreciated. Early on, the IMechE secretary was having to explain, in the pages of Chartered Mechanical Engineer, that the CEI was neither a trade union nor a replacement for the individual institutions. Further explanations, sometimes couched in terms which indicated just a touch of asperity, seemed to be necessary with some regularity over the next 15 years. There were also, initially, occasional mutterings about whether some of the engineering institutions included in the CEI were fit to rank alongside the Big Three.

The intended role of the CEI was the "regularisation" of the profession. Several of the engineering institutions had been involved in a Commonwealth Engineering Conference in the mid 1950s which had agreed a definition of "professional engineer".

The definition said: "A professional engineer is competent by virtue of his fundamental education and training to apply the scientific method and outlook to the solution of problems, and to assume personal responsibility for the development and application of engineering science and techniques, especially in research, designing manufacturing, superintending and managing. His work is predominantly intellectual and varied and not of a routine mental or physical character, but requires the exercise of original thought and, if necessary, the responsibility for supervising the technical and administrative work of others."

What the definition did not do, of course, was to specify the educational attainments necessary to demonstrate competence. So an early aim of the CEI was to bring the processes whereby individuals qualified for membership of the different institutions into line with each other, so that a professional engineer who qualified through the IMechE would have broadly the same type of experience and qualifications as one qualifying through a different engineering institution. With more than 40 institutions, not all of them members of CEI and some of them demanding little in the way of intellectual rigour, this was no simple matter.

The eventual solution was to allow only some of the institutions – those with the more rigorous membership criteria, including the IMechE – to grant their members the new and universal Chartered Engineer status and the right to use the letters "CEng". Individual institutions' designations such as "Chartered Mechanical Engineer" lapsed when the new CEng came into effect in 1967, though they remain in colloquial use to this day, and the IMechE did not rename its monthly magazine until the late 1980s.

A second and parallel role of the CEI was to set up the first-ever register of professional engineers. Again, the idea of registration proved controversial, and some members of the IMechE argued that engineers should not be encouraged to see themselves in the same professional light as doctors and lawyers, where a register was necessary to ensure fitness to practise. By contrast, the Labour government of the mid-1960s was keen on the register, and at one stage appeared to be volunteering to finance both the register and the computer system to operate it.

One consequence of the various CEI moves was that all corporate Members of the IMechE, to follow the nomenclature used in other institutions, became Fellows, and all Associate Members were translated into Members. Even this change, apparently welcomed by the overwhelming majority of IMechE members, was only achieved after an expensive postal ballot because seven of the 137 members who turned up to the special general meeting to approve the move called for a ballot, and under the democratic rules of the IMechE, that was enough.

Behind the controversy lay the long-running debate about the balance within the membership of the Institution of Mechanical Engineers between craft and science, and between the academic and the practical traditions.

The non-academic tradition within the IMechE was a strong one and, more than a century on, and with engineering firmly established as a subject of academic study, there were still members keen to invoke the spirit of George Stephenson and 1847 as justification for retention of non-academic routes to memberships. But by the mid-1960s, the pressure that the non-academics could exert was dwindling as the growth of university and college courses meant a more widespread availability of engineering qualifications. And all the engineering institutions had to be impressed by a manpower survey carried out by the Ministry of Technology, which indicated that large numbers of engineering graduates regarded institution membership as unnecessary.

The survey was a significant factor in propelling the CEI forward in its proposals to regularise the profession and to upgrade the academic requirements for institution membership. The survey finding suggested for the first time that the engineering institutions risked being sidelined unless they consciously brought their membership requirements into line with the wider education system.

With the phasing out of the old Higher National Certificates and the new Higher National Diplomas as the basic requirement for institution membership, the new members for all the institutions would be graduates or would have qualifications meant to be equivalent to degrees. And if engineering graduates were not convinced of the necessity of joining, the outlook for the institutions was bleak.

Having been criticised in its early years for inaction, the CEI moved smartly in the mid-1960s to set the new pattern for institutional membership. In parallel, the IMechE and the other individual institutions started a subtle policy of wooing engineering employers in industry: if employers could be persuaded that a graduate engineer with institution membership was a better option than a mere graduate, then the threat could be averted. At the same time, the IMechE set up a Graduates' Liaison Committee so that views of younger members

Factory visits and exhibitions formed an increasing part of the IMechE's programme in postwar years

could be conveyed to the Institution council, whose members still tended to be senior engineers of advanced years.

One effect of the change to an all-chartered status for the IMechE was to leave a gap in the institutional framework: there was no engineering institution for technician engineers and engineering technicians with mechanical engineering qualifications below degree level. These people in the past would have been able to qualify for IMechE membership but were now, as the educational requirements for membership were tightened across the 1970s, facing the probability of being disallowed membership for want of a degree.

The gap was eventually filled with the formation of the IMechE-sponsored Institution of Technician Engineers in Mechanical Engineering (ITEME), now the Institution of Mechanical Incorporated Engineers – though this was, of course, a further fragmentation within the profession as a whole and disunity was already cause for external and internal criticism.

All of these reorganisations and realignments took up vast amounts of time and energy. If there is a common phrase used in the annual reports to the Institution by the succession of presidents across the 1960s and 1970s, it is "two steps forward, one step back". At times, it must have taken supreme optimism not to have written the reverse: "one step forward, two steps back". The Institution was – and remains – resolutely democratic; the price of democracy is often that decision-making is not exactly speedy.

But if professional and organisational matters were the source of frequent headaches for the Institution, its members and its staff, there was also considerable satisfaction to be gained from the new sense of direction, starting from the early 1960s, as it applied to the Institution's technical and innovation activities.

The 1960s were years of massive technological confidence. There seemed no limit to the ability of science and engineering to deliver the technologies and the products that would enable mankind to

An image that gripped a generation: man's first steps into space suggested there were no barriers that science and technology would not eventually overcome

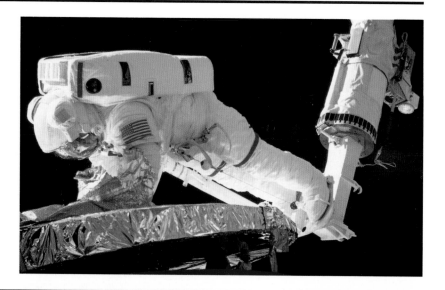

The 1950s and 1960s saw huge increases in car ownership, with mass-produced cars incorporating a new degree of stylishness as well as functionality

meet all its ambitions. Already in the 20th century, the last great inaccessible parts of the planet, the skies and the seas, had been explored with the development of the aeroplane and the submarine. Now, with the Americans and the Russians sending rockets into outer space, science and engineering appeared to offer the potential for man to break free from earthbound restraints.

Mechanisation had also made the leap from the factory into the home, and domestic appliances and private cars gave tantalising

Cars were being deliberately marketed to new segments of the population in the 1960s. Whether safe driving was possible in the shoe fashions of the time was another matter entirely...

promise of a world freed from drudgery and able to enjoy increased amounts of leisure time. Suddenly the products of technology were everywhere and technology was fashionable. The twin-tub washing machine and the electric iron reduced washday drudgery; then the drip-dry nylon shirt, product of materials technology, made ironing unnecessary.

Technology was a powerful and liberating force, and though the development of atomic weapons indicated that it was a force which could be used for destruction as well as for progress, the overall mood in the 1960s was positive and optimistic. It was, after all, the decade of *Star Trek* and of the gadgetry that helped James Bond win through.

In Britain, faced with the end of empire and declining status as a world power, there were maybe fewer opportunities for grandiose and ambitious projects on a world-beating scale, but technology and innovation were very much seen as forces to be harnessed to achieve economic prosperity.

Engineering on the heroic scale: prototype and preproduction models of the Anglo-French Concorde take shape at the British Aircraft Corporation, Filton

Almost 30 years after the first flight, Concorde remains the world's most unmistakable aircraft

And there were still a few projects of heroic proportions, of course. Concorde, the Anglo-French project to create the world's first supersonic passenger aircraft, was always economically doubtful, but it went ahead, surviving several close calls when spending was being drastically cut back. Its survival was due, to a large extent, to the strength of its prestige status as proof that Europe could compete on technology terms with the big battalions of the United States and the Soviet Union.

Perhaps of longer term significance, certainly in terms of overseas earning power, was the project to create the Harrier, the world's first short and vertical take-off and landing aircraft. As with Concorde, the

Concorde Front Fuselage production

Harrier came very close to cancellation in several of the government's periodic public spending purges in the early 1960s. It too survived, to become possibly Britain's most successful defence export in the last third of the 20th century.

Not all the major projects were related to exports and new technologies. At home, in perhaps the most concentrated burst of infrastructure development of them all, the whole of the national gas transmission network that had been laid down over the previous century and longer was upgraded in a dozen years to new standards required for the high pressure transmission of natural gas from the North Sea.

Changing landscapes: Bacton, in rural Norfolk, became the landing point for much of the North Sea's natural gas supplies

But an awful lot of the emphasis on technology, both in British industry and reflected in the meetings and conferences at the Institution of Mechanical Engineers, was to do with the vital but less glamorous end of innovation: increasing industrial efficiency by process and product improvement.

There was a degree of necessity about this. Though standards of living in Britain had risen steadily across the 1950s, domestic prosperity had been achieved to a large extent without regaining industry's international competitiveness. British companies had increasingly lost out to firms from the US and Europe in traditional markets in the Commonwealth, and in the 1960s the industrial development of countries such as Japan was creating new competition.

British industry faced particular problems with its legacy of old facilities and machinery, and its companies appeared at times to have difficulty with the basics of successful business. An example of the fundamentals going astray was the Mini, the revolutionary small car introduced by the British Motor Corporation in the late 1950s. The Mini, which crammed an awful lot of room into an awfully small space mainly by dint of having its engine mounted transversely, should have been a world-beater. Indeed, it did sell well in several overseas markets as well as at home, but BMC's "competitive" pricing policy meant that it never made sufficient money to generate the investment needed to keep the car ahead of its rivals or to allow BMC

Innovation in design: the first Mini was a revolution in automotive engineering and, in the 1960s, it was high fashion too

Minis rolled off the production line at Longbridge in huge numbers, but profit levels on the car were rarely sufficient to fund investment in a wider range of models

to develop a range of similarly competitive vehicles in different sectors of the car market.

With its newly-acquired industrial focus, the IMechE was active in early attempts to create by planning the machinery that might assist manufacturing industry to be more competitive. In this respect, it was, as ever, well in tune with the times. The national political consensus was that rational debate and planning were the methods by which better decision-making would be achieved. The Institution's willingness to become involved in the planning process and in the huge number of new organisations, committees and councils set up in the 1960s took it into new areas of influence.

In fact, in taking a more active line, the Institution was doing exactly what the politicians, of all major parties, were wanting. Lord Hailsham was Lord President of the Council and Minister for Science in Harold Macmillan's government in 1963 when he opened the conference held jointly with the Institution of Production Engineers to mark National Productivity Year. "Whichever way you look at it," he said, "the position of the engineer is critical for the future of Britain. This being so, it is a pity that there are so few of him (and relatively none of her) and that he occupies a status too far below the salt."

Hailsham went on to make the first pronouncement of what would become a regular government chorus in years to come, arguing for a simplification of engineering institutions and degree courses.

But it was a politician of a different colour, Harold Wilson, who made the biggest capital out of the popular appeal of science and technology. Portraying himself and the Labour party as modernisers, his first speech as Labour leader to the 1963 party conference put science and state-sponsored research as a cornerstone of a future Labour government's economic policy. He spoke of a scientific revolution, in whose "white heat" a new Britain would be forged with new technologically-led industries. The "white heat of the technological revolution" – the exact phrase was never, in fact, used by Wilson – became a Labour catchphrase, and when the Conservatives in the 1964 election campaign referred to Wilson as "a slick salesman of synthetic science" it reinforced the impression that technology was recognised on all sides as an important national issue.

In the end, of course, the catchphrase came to haunt the Labour government. The new Ministry of Technology (MinTech) set up after the 1964 election victory was intended to deal with the planning of industrial and scientific research, and the energetic Anthony Wedgwood Benn was later appointed to head it.

But the intractability of Britain's economic problems and the inefficiency of much of British industry proved greater hurdles than the modernisers ever suspected. By 1970, when MinTech disappeared into the new Department of Trade and Industry, the interventionist technology policy which was intended to lead and shape industrial research and innovation had been pretty much subsumed into an interventionist industry policy. And this, for governments of both main parties across the 1970s, all too often meant attempting to manage the decline of traditional industries rather than developing new industries or technologies.

Nevertheless, the processes of planning which were intended to help pick the winners for the future demanded expertise, and the IMechE was well-placed to help. In the early 1960s, it worked with government on establishing a committee on industrial design, from which the Design Council emerged. Within the CEI, it set up and provided the administrative support for a committee on technology and innovation which produced a succession of reports on product and process development.

The IMechE was also the means by which a whole new branch of engineering, with a brand new name, became recognised and established. The Lubrication and Wear Group was not one of the bigger specialist groups within the Institution, but the subject had a fairly long pedigree, dating back to Beauchamp Tower's research in the 1880s. In 1937, it had co-ordinated 53 scientific and technical bodies in the first-ever conference – called a "general discussion" – on lubrication and lubricants, with 136 technical papers and an attendance of 600 engineers; 20 years later in 1957, it repeated the event, with more emphasis on the friction and wear elements. The IMechE specialist group started holding annual conventions, and the subject also came

A new engineering science: tribology was chosen as the name for the increasingly important study of friction and lubrication in the mid 1960s

This has led to the recognition by a considerable body of opinion, including some of the leading experts in the field, of the limitations of the present term.

20. To remove the cause of the misunderstandings encountered and their consequences, and after consultation with the English dictionary department of the Oxford University Press, it is recommended that the term TRIBOLOGY* (triboscience and tribotechnology), be used for the description of the subject matter, with such subsections as appropriate, e.g. tribophysics, tribometallurgy, triboengineering, tribo-organisation, etc. The term 'lubrication' can then be used to describe the popular activities normally understood thereby.

21. Tribology is defined as follows :

TRIBOLOGY

is the SCIENCE and TECHNOLOGY of interacting surfaces in relative motion and of the practices related thereto.

This term includes the subjects of *friction, lubrication* and *wear*, e.g. :

The physics, chemistry, mechanics and metallurgy of interacting surfaces in relative motion including the phenomena of friction and of wear.

Fluid film lubrication, e.g. hydrostatic, hydrodynamic, aerostatic and aerodynamic.

Lubrication other than fluid film, e.g. boundary and solid lubrication.

Lubrication in special conditions, e.g. during metal deformation and cutting processes.

The properties and operational behaviour of bearing materials.

The engineering of bearings and bearing surfaces (e.g. plain and rolling bearings, piston rings, machine slides, gear teeth, etc.) including their design, manufacture and operation.

The engineering of bearing environments.

The properties and operational behaviour of fluid, semi-fluid, gaseous and solid lubricants and of allied materials.

The quality control and inspection of lubricants.

The handling, dispensing and application of lubricants.

The management and organisation of lubrication.

up at a lot of industrial administration and engineering production group meetings in the early 1960s, as well as at outside gatherings.

A point which emerged strongly was that there was a gap between the knowledge of the engineers researching into bearings and lubrication, often in universities, and the knowledge of the engineers operating plant in industry who were experiencing the real problems with bearings and lubrication.

At the instigation of the Department of Education and Science, Peter Jost was asked to form a working group to report on the wider issues. Naturally, many of the members of the working group were also members of the IMechE. The group did more than just report: it decided that there had to be a new and less clumsy name for "the study of lubrication, friction and wear". A new name, it was felt, would cement in engineers' minds the fact that this was a homogeneous subject, not several separate disciplines. The group consulted the Oxford University Press's dictionary department, and came up with "tribology", from the Greek word τριβoς (tribos), meaning rubbing.

So when, in August 1966, the Minister of Technology Tony Benn announced the formal setting up of a committee, again chaired by Jost, to implement the group's recommendations, he was able to give a rather snappier title than he might have done a year or so earlier. The Committee on Tribology operated from 1966 to 1972, and by the time its work was done, the term had been widely accepted around the world as the name of a branch of engineering which had existed for many years before, but which now had a new and necessary cohesion.

The Design Council work and the development of tribology illustrate the Institution's readiness to take an active role on national issues. It was now more than willing to do this, and to oversee activities that involved non-members as well as members. The change was in line with a new view of the nature of technology, and the IMechE was very much involved in developing this new view.

The new view was shifting the emphasis from invention to innovation. It was a subtle change, but an important one: it meant giving attention to areas such as design, manufacture and automation, and to factors such as efficiency, quality, reliability, longevity and maintenance. All of these subjects applied to the whole of engineering, from big projects to small components, and they linked into wider business concepts such as cost control and marketing. Tribology fitted into this pattern as an enabling or process technology, as did another new 1960s "-ology", terotechnology, the study of maintenance engineering.

The old idea was that competitiveness was gained by being first with a new product or a new technology; gradually, engineers were discovering that being first was less important than being better. It was a lesson that much of British engineering industry took many years to learn, and the learning process for many companies and individuals was painful.

Many of the "inventions" of the 1960s and 1970s were in fact also to do with process improvement rather than new products. And many of them were dependent on the ever-increasing availability and power of computers and on the development of control theory. Computer numerical control for machine tools and computer-based process control both date from before the 1960s, though neither was commonplace on the shopfloors of industry. The developments of the 1960s and the 1970s were in extending the range of activities where computers were involved, in bringing computing into the broad enabling technologies of design, manufacture and control and in linking engineering functions through to wider business applications.

The first computer-aided design system was developed at the Massachusetts Institute of Technology in 1963. Called Sketchpad, it consisted of a cathode-ray oscilloscope driven by a Lincoln TX2 computer, and displayed graphical information on the screen. Pictures would be drawn on to the screen and then manipulated using a light pen. By the beginning of the 1970s, there were maybe 200 CAD workstations in industry worldwide, most of them in aerospace and automotive companies, and most still used simply as computer-based drawing boards. In Britain in the early days of CAD, there was a strong climate of opinion that it was more likely to have uses in architectural rather than engineering design: this was rapidly overturned as the potential for linking design and manufacture emerged and within little more than a decade engineers were impatient at the speed that the technology was evolving.

Robotics, the application of computing to work handling and placement, actually started before CAD, but got going more slowly. The first industrial "robot" was probably the multipurpose manipulator patented in the US in 1954 by George Devol. It had a playback memory and point-to-point control. Devol later joined forces with Joe Engelberger and the Unimation company was formed, launching its first robot in 1962 after some delay because of the difficulty of developing the control system. The first industrial robot was used to unload a diecasting machine at General Motors' factory at Trenton in New Jersey. Within a few years, units could be programmed to do several tasks, and continuous path control had been developed.

For some time, though, robotics was pretty much a technology in search of an application. It was hampered initially by the limited numbers of computers in industry and, even when the coming of large scale integration and the microprocessor reduced the cost of computing power drastically, it was held back by the range of tasks which could be done by robotic means. Where numerically controlled machine tools replaced a skilled machine operator, industrial robots were, in their early days, far too intellectually challenged to be suitable for anything but unskilled labouring work, which made economic justification difficult. And if a company had paid out a huge sum of money for its first computer system, it was unlikely to see the range of tasks an industrial robot could perform as the best way of recouping the investment. Why buy a clever machine for an unclever job?

Several factors changed the perception of robotics. One was the development of systems and surroundings that allowed for the lack of sophistication of the robots; an example was the car body spray booth. Another was the use of robots for jobs which were deemed to be too hazardous for human beings, or which were downright unpleasant: this stimulus for robotics was helped by the health and safety legislation of the mid-1970s. And the economics changed: unskilled labour costs raced ahead of inflation for most of the 1960s and 1970s, whereas the costs of computer control systems plummeted.

But the factor which changed the outlook for both computer-aided design and robotics was the coming together of the technologies. Three-dimensional CAD, with solid modelling, was integrated with the automation technologies of robotics, machine tool control, and process planning and control within powerful but inexpensive computer systems based on the silicon chip. This revolution in manufacturing technology had started by the end of the 1970s, though it would be the 1980s before its use became at all widespread and before other technologies such as telecommunications would be integrated.

The idea of control of manufacturing operations and other branches of engineering was very much in keeping with the new mood of the 1970s. The first manned lunar landing in the summer of 1969 was the ultimate technical triumph of the technology-driven 1960s: there appeared to be no limits to what technology could achieve. Yet within four years, the world economic order had been turned upside-down, and expansionist optimism had been replaced by introspection. Good housekeeping, or cost control, became the most prized industrial virtue as money suddenly got tight.

In Britain there was a succession of blows. Early in 1971, Rolls-Royce, synonymous with engineering excellence and luxury, collapsed. The reason seemed to be the amount of money which had

Mission accomplished: the first test flight for the RB-211 engine was a 1960s media event

Lord Stokes proposed the concept of engineering accountability in his IMechE presidential address. Many held him accountable at least in part for the long-running woes at British Leyland

been consumed developing the new range of RB-211 aero-engines and particularly the carbon fibre fan blades. The collapse made unhappy reading for engineers. The RB-211 was a technology triumph in almost every respect except one: there was a strong implication that the technology had been pursued with little thought given to the costs. In some quarters, it was openly asked whether Rolls-Royce did not prove that engineers were unsuited to top jobs running big companies.

Rolls-Royce was a dramatic illustration of a point the IMechE had made many times over the preceding 50 years: that engineering excellence could not be achieved at the expense of industrial success. But this was not the only famous British engineering name in trouble in the early 1970s. Across the 1950s and 1960s, the British motor industry – or at least most of the British-owned part of it – had gradually come together into one enormous company, British Leyland.

As an agglomeration of umpteen different companies with a huge mostly unrationalised model range, large numbers of factories, many unhelpful working practices and the burden of national expectation, British Leyland was always likely to be a difficult beast to control. In world motor industry terms, it was a small player in a league increasingly dependent on international muscle; in business terms, it had inherited a range of problems from companies which had banded together largely because they were not strong enough to stand on their own; in terms of investment and technology, the success of the Mini had not been built on. A succession of financial crises at British Leyland culminated, in the mid-1970s, in the British Government bailing the group out by taking it into public ownership, as Rolls-Royce had already been.

Ironically, Lord Stokes, the British Leyland chief executive, was president of the IMechE in 1972, just at the time when the first crises of confidence across manufacturing industry were starting to be felt. In a presidential address remarkable for its candour about BL's problems, Stokes introduced a concept which would recur across the

The first Rolls-Royce RB-211 aero-engine represented progress in engineering technology on several fronts, including the carbon fibre blades. But it contributed to the company's downfall

years: engineers, he said, had to be accountable for their actions. Engineering was not an activity carried out in isolation from the rest of industry or the rest of a company.

"If an engineer is to make a really effective contribution to this industry he must be able to stand back from his immediate task, be it of design, development or management, and to place it in the context of the enterprise as a whole and the objectives to which it is working," he said. In an age of ever-increasing specialisation of engineering skills, engineers had to be generalists too, to be able to see the wider context.

Meanwhile, out in the wider context, other industries too were falling into difficulties. Many of the problems were exacerbated by the world energy crisis triggered by the huge increase in the price of oil in 1973. Markets such as shipbuilding collapsed virtually overnight; with so many of its customers reining back on orders, steel-making in Britain moved sharply into losses. The energy crisis was followed by a series of economic crises as inflation in Britain hit record levels and sterling slumped.

At the Institution of Mechanical Engineers, inflation became a recurring theme, largely in terms of regular meetings to raise subscription levels. At the end of 1975, the members rebelled, rejecting the council's proposals for yet another increase; with inflation running at 30 per cent and more, warnings from the president and the secretary that the consequences could be a severe reduction in activity, or even putting the Institution into financial difficulties, were not idle threats. The increase in subs was resubmitted to the members, and approved.

The old steelmaking methods shown here disappeared in rationalisations and modernisations across the 1950s and 1960s. But a more modern industry in the 1970s faced different problems

It was not all gloom, of course. Natural gas transformed the engineering and business prospects for the gas supply industry, a sector that in some parts of Britain appeared to be in terminal decline. Then the first crude oil from the British sector of the North Sea was pumped ashore in June 1975.

Offshore oil and gas development offered the potential for a whole new engineering industry, as well as the possibility of reviving those heavy engineering sectors which had been so badly hit by the growth of worldwide competition from low-cost developing countries. The developments certainly changed the industrial geography of Britain, bringing multinational process engineering companies to the highlands of Scotland and paving the way for the huge inward investment into Britain in the 1980s.

There were technical triumphs in the North Sea development too. Huge rigs were built in converted shipyards and in purpose built facilities along the east coast; offshore platforms and onshore oil and gas reception plants gave a fillip to industries such as instrumentation.

Yet there was a public perception that British engineering industry was not taking full advantage of the opportunities of North Sea oil and the perception was not wholly unjustified. For example, the industry could not supply the volume, and to some degree the quality, of steel required both for pipelines and for steel jackets. British industry proved unwilling to invest in the pipelaying barges involved in the essential infrastructure of hundreds of miles of subsea pipeline.

Huge structures of a kind not made in British engineering industry before were called for in the development of the UK North Sea oilfields. This is the Graythorp I platform for BP's Forties field

The international oil industry also had a long-established supply chain based predominantly in the United States and to a lesser extent in the Netherlands, France and Germany. The pace of development was so great that there was no easy introduction to the market for UK companies, and government assistance was needed to ensure that British suppliers were able to make an impact. Moreover, the deep and hostile waters of the northern North Sea required technology advances which were far from easy for established offshore suppliers, let alone for British newcomers.

In fact, Britain and British engineering did well out of offshore development, and the North Sea certainly cushioned the effects of industrial decline in some important sectors as well as generating important revenues for the exchequer. But the rise of the new offshore business could only offset partially the crisis of confidence elsewhere in British industry in the 1970s.

The effect on engineers of the series of business and industrial failures was to reinforce the need for pragmatism in engineering. Technological flamboyance had been the undoing of Rolls-Royce;

Uncharted waters: The North Sea yielded first natural gas and then oil. Here, BP's Sea Quest drilling platform prospects 110 miles off Aberdeen in 1971

lack of focus had been a factor in British Leyland's problems; economics, both the national economy and in terms of the health of individual companies, could no longer be ignored by engineers. The need across industry was to get more from less, and the challenge especially to mechanical engineers was to use innovative technology to improve industrial processes and methodologies.

The challenge created some resentments. To a large degree, engineers saw themselves as being asked to do what they were already doing and as being blamed for wider failures in corporate governance in Britain. Japan was widely held up as the example of a successful manufacturing economy, yet the engineering factors which had contributed to Japanese success seemed oddly familiar to British engineers: less familiar were the commitments to investment in product and process research and in shopfloor technology, the emphasis on the customer's business needs for quality and reliability of service, and the close relationship between industry and government. On all of these factors, there was evidence across the 1970s and into the 1980s that engineers felt let down by British industry and non-engineering management.

The resentments surfaced many times at Institution of Mechanical Engineers meetings and in the Institution's magazines. Complaints about the status of engineers had been heard intermittently virtually

In little over 20 years, a whole new industry was born, incorporating leading-edge technologies from across the range of engineering. This is Shell's Gannet field on a calm North Sea day in 1993

The North Sea's technology advances required the development of new engineering skills in new and challenging locations

since the Institution was founded; but the chorus of bitterness grew sharply through the 1970s. Surveys of engineers showed much job satisfaction in terms of engineering, but much job dissatisfaction about the way engineers were treated within companies, and as professionals. "Underpaid and undervalued" summed up the attitude of many.

The tensions affected the unity of the Council of Engineering Institutions, and at one stage in the mid 1970s the Institution of Electrical Engineers threatened to withdraw. In fact, the CEI suffered continuously from a lack of clarity about its role; many individual institution members seemed unclear about what it did, and at times the institutions which were meant to be co-ordinated by the CEI seemed to disagree about the nature and degree of co-ordination required.

In the late 1960s, with the register of engineers under way and all institutions bringing membership requirements into line, there were serious suggestions that the CEI might be the basis for a wider Institution of Engineers. The suggestions coincided with grand plans for large-scale redevelopment of parts of Westminster: one idea was to build a CEI Centre – a tower block, naturally, given the prevailing architectural fashion – on the site of the Institution of Civil Engineers right next to the Mechanicals' headquarters. The Centre would house the new all-embracing Institution, plus any of the sectional engineering institutions that cared to survive the amalgamation. It was almost a re-run of Sir William Siemens' plans of the early 1880s.

Nothing came of this plan – as nothing had come of Siemens' – and across the 1970s nothing much came of other plans to upgrade the CEI or give it new direction. There were other plans too to bring individual institutions together. At one stage in 1972/73, the IMechE was involved in separate negotiations with the Institutions of Electrical Engineers and Production Engineers about "closer collaboration".

Britain's own contribution to Japanese engineering success was often overlooked. The first post-war Nissan car was the Austin A40 Somerset, handed over in the 1950s

The to-ing and fro-ing between the different institutions had the beneficial effect of putting in place a growing number of joint activities and conferences and forging links between groups within the different institutions.

Even when the ideas for progress came from the presidents of the three biggest institutions, the result was little in the way of action. In 1974, the presidents of the Mechanicals, Electricals and Civils put forward a plan for a restructuring of the profession into a new Institution of Engineers based on the register in which individual engineers would be the members. The plan would leave the existing institutions intact to pursue their sectors of engineering. As an idea, it was politically astute, well connected – yet it got nowhere.

A couple of years later, the three presidents' plan was still slowly passing through CEI subcommittees and the individual institutions. The IMechE and the Electricals then came up with yet another scheme: a Corporation of Chartered Engineers along the lines of the General Medical Council. That too made little headway.

The CEI scheme which did make progress was the setting up, in 1976, of the Fellowship of Engineering, a deliberately elite body composed of 600 of the most distinguished engineers from all branches of

The man of the report: the independent-minded former British Steel chairman Sir Monty Finniston was commissioned to report on the engineering profession

the profession. Prince Philip gave strong backing to the initiative and became the senior fellow; Lord Hinton of Bankside was the first president. Membership of the fellowship allowed chartered engineers to put the letters FEng after their names; it was, and remains under its present name of the Royal Academy of Engineering, the ultimate accolade for individual engineers.

With little progress in prospect on the issues affecting the engineering profession as a whole, the IMechE – and the other big institutions – turned its attention very much to ensuring its own relevance to its individual members and the industries they were working in.

A consistent and probably unnecessary worry for the IMechE across the 1970s was membership. Having passed through the 70,000 mark by the end of the 1960s, membership peaked in the early 1970s, and then started to drift downwards – not by a lot each year, but after nearly 30 years of staggering growth, the change was unexpected. The group structure again came in for examination, and the range of services for members was looked at. The aim, said the working party formed to explore future services and activities, should be "more broad technical help to the average engineer throughout his professional career".

Another substantial worry was finance. With its charitable status under the charter, the Institution was not in the business of making profits, but neither did it have the reserves to be able to absorb consistent losses. In the mid-1970s, inflation ate away at steady sources of income such as rents from properties and other investments, and businesses such as publishing and conferences proved highly responsive to national stop-go economic policies. Members' subscriptions accounted for less than half the total income.

Part of the reduction in the Institution's cost base which was put in train in response to the financial problems was the relocation of some of the headquarters functions, including much of the publishing business, out of expensive central London. After a long trawl of potential sites, the IMechE set up its second satellite offices in Bury St Edmunds.

The IMechE's own difficulties were part of a general malaise in engineering both as an industry and as a profession and the malaise did not go unnoticed in high places. In 1977, the government appointed a Committee of Inquiry into the Engineering Profession under the chairmanship of the former British Steel chairman Sir Monty Finniston. The Finniston committee's brief was a wide one: it was to look across the whole of manufacturing industry, the education system and the bodies that regulated the engineering profession; it was encouraged to investigate international performance in terms of industry and engineering; and there was at least a tacit understanding that the Labour government which had commissioned the report would be bound to act on its recommendations.

The engineering institutions made their views known singly and collectively through the CEI. There was a natural degree of wariness, since a premise behind the establishment of the Finniston committee appeared to be that the present self-regulation arrangements in the engineering profession had not worked well.

With proposals for the revamp of the existing CEI on the table – the proposals were a legacy of the three presidents' initiative of 1974 – there was some feeling that the government had jumped in just at the point where self-regulation was about to prove itself. In addition, many engineers felt that the achievements of the CEI in bringing engineering qualifications into line and in setting up the register had been understated and there was widespread disquiet about the implicit criticism of the profession. Other professions such as medicine and the law regulated themselves without government interference. Why did engineers have to put up with this kind of scrutiny?

Alongside the wariness, though, was a general recognition that individual engineers were keen to see some improvement in their status, and some relief that the inquiry's terms of reference explicitly acknowledged the central importance of engineers and engineering to the economy.

The Institution of Mechanical Engineers was the first of the engineering institutions out of the blocks in making a submission to the Finniston inquiry: in fact, its evidence arrived before the membership of the inquiry committee had been announced. It argued that engineering should not be blamed for the malaise in British industry, that employers should not expect "ready-made" engineers to emerge from the education system and that the five-year "Chartered Engineer package" of education and in-company training was the best way to produce competent professional engineers. It came down in favour of a statutory register of engineers to be run by the CEI, but added that it was vital that the profession should be self-regulating and "save for matters of public interest, free from outside interference".

A few weeks later, the IMechE came back for a second submission. It focused on two aspects of engineering education: the need for management and leadership training for engineers and the desirability of schemes that would allow technician engineers to progress to chartered engineer status.

Industry Secretary Eric Varley, who had commissioned the Finniston report, had asked for a quick inquiry. In the event, Finniston and his colleagues had to wade through 700 separate submissions, 200 of them from institutions, associations and companies, 50 meetings to discuss individual submissions, a hundred or so working group meetings to discuss particular aspects of the inquiry, and 30 plenary sessions of the whole committee. In these circumstances, to produce a report in just over two years was not at all unreasonable. It was, though, too late for Varley: in May 1979, he and his Labour colleagues were voted out of office and it was on to the desk of his successor, the arch non-interventionist Sir Keith Joseph, that the report eventually dropped in January 1980.

The change of government, and particularly the appointment of Joseph as Industry Secretary, gave the Finniston report severe problems. The report argued for the creation of a statutory Engineering Authority, appointed and financed by the Industry Secretary. The new Authority would have powers to regulate the profession including approving university courses and running the professional register. The institutions came in for only a little criticism, but the report suggested that their strength lay in their learned society activities, and that they should be encouraged to stick to those in future. The CEI was more harshly treated: while recognising its "not inconsiderable" achievements since 1965, the report said it "has not made a significant impact upon the fundamental problems of establishing greater understanding of the nature and role of engineering and in promoting the engineering dimension in national economic affairs".

Much of the Finniston report echoed what the IMechE and many of its members had been saying for a long time. There was strong support for the central thesis that engineering was vital to the welfare of industry and the economy as a whole; the concept of engineering professionalism was supported; the report was keen on longer and broader engineering degrees, which had been advocated in several institutions.

The IMechE council, however, reacted with uncharacteristic savagery. The report, it said, contained "fundamental errors of method and judgement". The range of tasks allocated to the new Engineering Authority was "so unrealistic as to border upon fantasy".

The Institution went on: "It recognises the importance of an 'engineering dimension' and accepts that in successful companies that dimension is well in evidence, but instead of addressing itself directly to ways of implanting the necessary awareness in those parts of industry that lack an engineering dimension, it calls for a drastic and needless change in the organisation of the engineering profession which has probably done more to make an engineering dimension possible than any other element of industry."

The engineering education proposals came in for particular criticism. The inquiry "has made proposals for new standards of education, training and individual accreditation without having informed itself adequately on the level of the existing standards. The result is that the criteria they set for registration are inferior to those currently applied."

The IMechE was not alone in criticising the Finniston report. In a personal comment, Lord Hinton of Bankside attacked the inference that engineers were in any way responsible for Britain's poor industrial performance. By and large, he said, British engineers were outstandingly good; the problem was that not enough people outside the engineering profession appreciated the fact.

The Finniston report's remedies as well as its analysis came under fire. Virtually all the engineering institutions viewed the idea of a statutory authority with suspicion and wondered publicly about how they were meant to operate alongside a body over which they had no control and only limited influence. In so doing, the institutions took quite a lot of flak, some from members and some from outside commentators, about what could be seen as "sour grapes". But while there may have some elements of resentment that Finniston was

United they (finally) stand: Sir John Fairclough (right) renegotiated the role of the Engineering Council with the institutions. Here he hands over the Royal Charter of the Council to his successor, Dr Alan Rudge

aiming to reduce the power and influence of the institutions, behind the criticism of the report was genuine concern that an opportunity to tackle a fundamental weakness in British industry had been missed.

In the end, of course, the fact that the Finniston committee reported to Sir Keith Joseph, and not to a more interventionist minister, proved virtually as important as the institutions' objections. Joseph was not in the business of adding to industry's legislative burden if at all possible, and was strongly opposed to the creation of any more quangos, or non-elected bodies.

The Engineering Council which he eventually approved was a long way from the statutory body Finniston had proposed. The engineering institutions continued to lament the council's lack of democracy, and there were occasional tensions caused by overlapping functions – or, more frequently, by lack of communication.

But there was recognition – sometimes grudgingly given – that the Engineering Council did provide a focus and was able to bring a more heavyweight influence to bear on government departments dealing with industry and education than the old CEI had done. By the early 1990s, when the then Engineering Council chairman Sir John Fairclough sought to end the divisions between the council and the institutions once and for all, he found the institutions more than ready to talk. And the IMechE had used the intervening years to cement its relationships with industry and to complete its own transformation into a dynamic industry-oriented information-based business.

7 Recession and Revival

1980 TO 1996

Engineering has changed more in the past two decades than it did in all of the previous 130 years and the Institution of Mechanical Engineers has had to adapt to keep pace with the rapid change. Engineering is now a truly global activity, and engineering companies increasingly see the world as their market, using technologies and equipment which would be recognised in equivalent firms the whole world over.

In Britain, manufacturing industry and engineering industry in particular has appeared to regain a lot of the momentum it lost in the 1970s, and the engineering profession too seems to be more united and harmonious than for many years. As a whole, engineering has emerged from a long period of self-doubt and self-examination strengthened and with renewed confidence.

The changes of the past 20 years have been brought about partly through technology, and in particular the convergence of engineering disciplines which has resulted from the widespread application of computer-based and electronic systems. Partly, too, the changes are the result of changed world economic circumstances. Many of those

Converging tracks: mechanical engineering helps to build the computing technology that expands the role of mechanical engineers

circumstances have been adverse and have caused severe dislocation to individuals, companies, whole industries and, at times, whole national economies.

But to a large extent, the dominant change of the past two decades has been a subtle one involving engineering philosophy and culture to which all of these factors have contributed. Recession, the integration of engineering functions through computer-based technologies, new management ideas, particularly those developed in the Far East – all of these have helped to switch the focus of engineering innovation away from the technology of invention and towards the technology of efficiency. The need for competitiveness has become paramount, and engineers are at the heart of the drive towards efficiency. On the way, though, some of the faith in technology as a driving force of progress has been called into question.

The IMechE itself has been far from immune to change, and has also had its share of the hardships. With a large proportion of its business in publishing, the Institution corporately suffered from the economic rollercoaster of the 1980s and early 1990s. And many of its individual members, up to and including presidents, have known at first hand the insecurities of the changed employment market.

The past 20 years have also marked the final stages of the evolution of the Institution itself into a business-like organisation, and figuratively the culmination was perhaps the change in the title of the secretary: Richard Pike, the first-ever director-general (and secretary), was appointed in 1993, symbolising the completion of the transformation from technical gentleman's club to an association for professional engineers.

In practice, the evolutionary process within the Institution had been going on for the whole of the postwar period; there is little doubt, though, that for the IMechE, the end of the 1970s represented some form of watershed. And in many respects, by the end of the 1970s engineering as a whole, both the industry and the profession, were prepared for a period of upheaval. Nationally, the previous decade had been marked by continuing and well-meaning attempts to contain and control the effects of industrial and economic change.

Even before the first oil price shock in 1973, Britain's national competitiveness had weakened because of rapid monetary expansion and inflation, and the deterioration in the trade balance led in 1972 to the first of several sterling crises. When the Arab nations restricted oil supplies and raised prices in 1973, all Western economies were hit. But in Britain in the mid 1970s, the consequences were especially severe: in real terms, 1974's trade deficit was, in real terms, the worst ever and, in the following year, inflation was running at 15 per cent above the Western countries' average.

In these circumstances, the coming of North Sea oil into Britain's economic equations was something of a double-edged sword. The offshore oil allowed the trade deficit of £14 billion in 1974 to be turned around into a surplus of £5 billion by 1985. It provided a new engineering challenge and new business opportunities, and though British companies needed a little help to get started in an unfamiliar sector, there is little doubt that oil had a hugely beneficial effect on the trade balance.

But the oil also changed the international perception of Britain's prospects and for other parts of UK manufacturing industry the effect was not helpful. The pound had been allowed to float from 1972, and had fallen in value from $2.40 to $1.60 between 1972 and 1976. Britain's new status as an oil producer saw it rise back to $2.40 by 1980. The oil allowed Britain nationally to pay its bills, but individual companies and industries suffered.

Large industrial companies, many in engineering sectors, found their markets changed and their international competitiveness suspect. Industrial relations were soured as resentments built up about the erosion of wages and living standards through inflation and about the loss of job security. The idea of progress through technology and innovation that had seemed to be a national priority in the 1960s dropped down the list; industry seemed stuck in a rut, unable to raise its performance, unable to generate much enthusiasm.

By 1979, there was a widespread feeling that Britain could not go on as it had been doing, and that radical and structural change could no longer be resisted. But as the new decade came, the chance that British industry and engineering could make an orderly transition to a new and more competitive basis disappeared virtually overnight. And many mechanical engineers in Britain found that, with devastating speed and frightening starkness, they were jolted into real fears for their jobs, in a lot of cases for the first time ever.

The recession of the early 1980s was a worldwide phenomenon and was brought about mainly by the second oil price shock: in Britain, though, the oil price rise was added to all the other stresses and strains within the economy as a whole. It was the final straw. The plunge into recession was sudden, dramatic and damaging.

In overall economic terms, the 1980s recession may not have been as deep as the worldwide depression of the 1930s, nor as long-lasting as the downturn at the beginning of the 1990s. Because of North Sea oil, there was the anomaly that the trade balance continued to move sharply in a favourable direction through the recession.

But for manufacturing industry in Britain, and for engineering industry in particular, this was almost certainly the most traumatic recession of them all. And the changes which it brought about in working practices, business organisation and technologies were in many respects as far-reaching as those that resulted from the first Industrial Revolution.

Before the gain, though, there was a lot of pain. Some of the casualties had been suffering for many years. In basic industries such as steel, shipbuilding, motor manufacture and across the range of heavy engineering, British companies had long been hard-pressed to compete against more flexible and lower cost rivals. In many of these sectors, technology was barely an issue, labour represented a large proportion of the total costs, and markets were increasingly global. Newly industrialised countries, particularly in the Far East, were able to take market shares in important sectors because of their low costs and their ability to deliver.

But it was not only in what were termed "sunset industries" that British firms suffered. Many appeared handicapped by history. They had ageing facilities, unrationalised production sites, unhelpful industrial relations and, in too many instances, a strange, otherworldly diffidence about the need to sell products and ideas. Some of the diffidence was alleged to be the result of an over-reliance on what used to be captive markets for British goods in the old Empire. But 30 years after the end of the Second World War this excuse for a lack of marketing skills and salesmanship was starting to wear a bit thin.

In addition, it had to be said that technology and innovation were commodities associated with relatively few British firms, and that

Traditional industries like steel were faced with stark choices: British Steel's modernisation across the 1980s turned it from the epitome of monolithic under-performance to the smartest steelmaker in Europe

much of British manufacturing industry was ill-equipped in terms of efficiency and flexibility to cope with the economic shock waves.

The Finniston inquiry into the engineering profession had been asked, as a secondary objective, to analyse Britain's engineering industry. Just before the Finniston report's publication, Sir Peter Carey, the permanent secretary at the Department of Industry, was invited to give the Thomas Hawksley lecture to the Institution of Mechanical Engineers. Hinting at what Finniston would have to say, Sir Peter did not mince his words: "The cause of our current economic weakness lies in the poor performance of our manufacturing industry," he said. A large part of the poor performance, he added, was due to the consistent under-use of engineers within British companies and a national under-valuation of engineering, both of which had to be rectified in the coming decade, the 1980s.

In the event, the understanding and support at the heart of government, which Sir Peter Carey's lecture implied, was suddenly of a rather different kind from the type shown by the successive interventionist governments – of both main political parties – in the previous 20 years. The new Conservative government elected under Margaret Thatcher in 1979 saw its prime task as putting the whole of the national economy on a sounder basis, even if that demanded some pain for individual companies and their employees.

If UK companies were losing out in world markets because of low productivity or because outdated technology made their products uncompetitive, then the government view, expounded most forcefully by industry secretary Sir Keith Joseph, was that it was up to the companies themselves to change matters. The government would be sympathetic, but it was not going to interfere with the workings of the markets.

The government's attitude to industry can be summed up in an exchange between a senior minister and a prominent industrialist which in fact never happened but which seemed nevertheless to pass instantly into engineering folklore.

At a time when it seemed to many that the recession in manufacturing industry would never end, the House of Lords' select committee on science and technology conducted an inquiry into the state of industry, taking evidence from both politicians and industrialists. The committee interviewed the then Chancellor of the Exchequer Nigel Lawson and heard also from Lord Weinstock, the chief executive of one of Britain's biggest engineering companies, GEC.

In the story, the Chancellor was asked whether manufacturing was not an essential part of the UK's mixed economy. He replied to the effect that there was nothing inherently special about manufacturing, and that there were other means of wealth-creation which could stand the British economy in good stead. To which Lord Weinstock was said to have retorted that the government appeared to want an economy founded on hamburger bars.

Throughout the early 1980s, though, there was a degree of ambivalence in engineering industry about the government's attitude and about the recession as a whole. The 1970s were increasingly seen as an era when issues had not been faced up to, and the results had been rampant inflation, dreadful industrial relations and declining competitiveness. At least some of the root problems were perhaps now being addressed, or being better understood.

The difficulty was in knowing where the process would end. Perhaps it was right that sectors where there was little prospect of Britain ever being able to compete against lower-cost rivals should be given up. But where should the line be drawn? At some stages during the recession, it seemed that no part of engineering industry in Britain was safe, either as a source of employment or as a British company: if it wasn't under threat of closure, it was very likely to be for sale. Even parts of the industry geared to defence, a relatively buoyant market under a government pursuing a pugnacious foreign policy, were not immune to the effects of rationalisation and international competition.

In the circumstances, it was not surprising that engineering, as both an industry and a profession, went through something of a crisis of confidence. The effect of the recession on large tracts of the engineering industry was shattering. Famous company names disap-

Defence industries, helped by wars in the Falklands and the Gulf, consistently did better than civilian industries in the 1980s

peared; employment in the industry fell drastically; factory closures blighted individual careers and whole communities.

For many professional engineers, the recession brought a new and unwelcome experience: job insecurity. Across the 1970s, the Institution of Mechanical Engineers had joined with other engineering institutions in producing regular surveys of members' pay and employment conditions. One consistent feature of the surveys was the degree to which engineering was a "safe" job: unemployment was barely known among professional engineers, and job mobility was far less for engineers than for many other professions.

The recession of the 1980s changed that for many engineers. Company closures and factory rationalisations put professional engineers out of work or on to the motorways as jobs shifted from one part of the country to another. The concept of lifetime employment for engineers within one company, which had been diminishing anyway, disappeared almost totally.

But curiously, while the ferment in engineering industry affected professional engineers' employment, it had little impact on their employability. The pay and employment surveys of the early 1980s – they were taken over by the Engineering Council – show that engineers were having to change jobs more frequently as companies went to the wall or "rationalised", but that qualified engineers rarely lingered long in the dole queues. When national unemployment soared to more than 10 per cent, the figure for professional engineers reached just 2 per cent, and many of those were genuinely "between jobs". Graduate engineers emerging from universities and polytechnics found that, while employers were no longer exactly queueing up for their services, they had much better job prospects than their fellow-students with degrees in virtually all other subjects.

Future fighter: the cost and timescale of aircraft developments such as Eurofighter made international collaboration essential

Some experienced engineers also took advantage of euphemisms such as "early retirement" and "career reorientation" to become consultants, contractors or new engineer-entrepreneurs. If insecurity was one of the features of the early 1980s, so too was a strange feeling of liberation for many.

What the recession was doing to manufacturing industry served to confirm the absolute importance of engineers and engineering, the message that Sir Peter Carey, Sir Monty Finniston and many other engineers had already emphasised. If shopfloor productivity or manufacturing efficiency were problems, only engineering would sort them out; if products needed updating, then only engineers could do it. The only way out of the recession for companies in industry was

To compete, British car companies had to adopt technologies such as robotic welding that had been pioneered overseas

Mass production 1980s style:
the Metro body welding line

for them to increase their competitiveness, and professional engineers were the key to that.

The result was that though overall employment in engineering industry fell dramatically in the years between 1979 and 1985, employment of professional engineers increased in the same period, not just as a percentage of those working industry, but in absolute terms. For the engineering profession, though, the recession of the early 1980s did more than just change employment patterns. It changed the profession itself.

Having gently and gradually moved, across 30 years, towards an all-graduate membership, the engineering institutions had to accommodate a new factor: that the pace of technology change was now so rapid that engineering graduates emerging from the education system would find their many of their skills and much of their knowledge out of date within a few years of graduation.

The Institution of Mechanical Engineers had always had an unofficial role in continuing engineering education. The various routes to qualification for membership had always assumed that engineering education continued into the workplace even for graduates, and the routine of learned society activities had involved, since Victorian times, the dissemination of information and intelligence about new techniques and technologies.

The long-awaited successor to the Mini, the Metro had a lot going for it. But the depredations of the UK motor industry in the 1970s meant that it struggled to compete with higher volume cars from companies with worldwide sales networks

Until the 1950s and 1960s, however, the learned society role, though recognised universally as important, was offered by the Institution and accepted by members on a fairly ad hoc basis. The Institution made no claim that its lectures, meetings and occasional conferences gave comprehensive coverage of the vast subject of mechanical engineering, and for their part the members involved themselves in Institution activities pretty much as it suited them. Once qualified, many members of the Institution played no further active part, and quite a few engineers who would have qualified for membership did not bother to join.

This was acceptable in an age when jobs were often held for life – or at least when many engineers stayed in a single company for the whole of a working life and so long as the companies did not specify the need for professional membership. It was also not a problem when the pace of technological change, particularly in manufacturing technology, was relatively slow-moving, and when engineering companies were to some extent at least protected from the harsh winds of international competition through captive markets. Mechanical engineers undeniably missed a lot if they decided not to join the IMechE or if, after joining, they failed to play an active part, but it would have been difficult to say that careers had been harmed by non-membership or non-participation.

But the post-war world was different. The pace of technological change was increasing; more than that, technology change was proceeding on a broader front, so that mechanical engineers needed knowledge of related disciplines – in electrical and electronic engineering, for example – to incorporate into new products and new techniques. On the business side, captive markets were disappearing and worldwide competition was developing.

In a climate where industry wanted more from its engineers, and engineers in turn needed institutions that would provide them with relevant and up-to-date information, engineering knowledge and engineering skills for both companies and individuals needed constant topping up.

The relaxed view of the educational role of the IMechE had begun to change with demands for greater industrial relevance from the late 1950s onwards, and the members responded to the changes in the Institution with increased numbers taking an active part.

The early 1980s completed this process. Taking their cue from the government and from the harsh economic realities, engineering companies were forced to face up to their own competitiveness, and the role of engineers was crucial for that. Employers needed to know that their engineers were abreast of the latest technologies, in terms of products, of course, but even more so in terms of underpinning technologies such as design and manufacturing techniques – and all the functions that were becoming computerised. The engineers themselves needed to be able to update themselves to retain their present jobs or to progress their careers. And the IMechE, as the unifying organisation covering all mechanical engineers, grabbed for itself the central role in providing the continuing professional development – CPD – demanded both by the individual engineers and their employers.

It was helped in this by the restructuring which started in the late 1970s and which was put into place just as the recession took hold in 1980.

The previous reorganisation of groups within the IMechE at the start of the 1960s had begun the process of linking the Institution more closely with the engineering industries where a large proportion of the members worked. A complex system of groups divided into horizontal "scientific" subjects applicable to most engineers and more specialised vertical industry-based subjects had been created; outside these groups, the Institution retained a separate Automobile Division, a legacy from the takeover of the Institution of Automobile Engineers soon after the Second World War, and had also created a separate Railway Division in 1969 when the Institution of Locomotive Engineers became part of the IMechE.

The new restructuring which took effect from May 1980 created broad divisions based on specific segments of engineering industry. Alongside the Automobile and Railway Divisions and the technology-based groups, new divisions were formed to serve power industries, process industries, engineering manufacturing industries and aerospace industries. Significantly, each included the word "industries" in its title and, considering the delicacy of relations with the other engineering institutions in the Finniston period, the aerospace industries division was only formed after consultation with the Royal Aeronautical Society. Later, an engineering management division was added.

The intention with the new divisions was that they should be able to work closely not just with the individual members employed in the industries, but also with the industries themselves. The aim was to create a form of mini-Institution with much of the responsibility and authority devolved to the divisions. Without breaking from the tradi-

The growing importance of aerospace as a high technology mechanical engineering business led to the formation of the aerospace industries division at the IMechE

tion of individual membership, the IMechE was broadening its relevance to bring in companies as well. And one of the areas where it could be of most use to both individual engineers and engineering companies was in updating skills.

There was, though, a further factor in the re-skilling of mechanical engineers and it too was brought into focus by the recession. Britain had long been a favoured European base for multinational companies, and many mechanical engineers worked for global companies, particularly in the international automotive and aerospace industries, and latterly in sectors associated with North Sea oil and gas.

One effect of the recession was to make Britain an even more attractive base for international firms to site their European operations. Britain's attractions were not necessarily to be regarded universally as plus-points: by comparison with other European countries, one advantage was that our costs, such as wages, were comparatively low. But our incentives were also generous and there was the abiding assistance of the English language, more and more the international language of business.

And equally, the existence of a long engineering tradition and the promise of up-to-date technical skills both contributed to the success of inward investment from the mid 1980s onwards. The international companies looking for European manufacturing bases were not willing to accept lower standards in terms of productivity or technology than they had in their home markets and the ability of British engineers and British workforces to deliver was crucial. Britain could offer good value for money, and engineering talent was an essential part of that good value.

The newcomers attracted to Britain as a European manufacturing base were not just in the sunrise industries of computing and electronics, though many were. The British car manufacturing industry was revitalised by the introduction of Nissan, Honda and Toyota, and "traditional" engineering sectors as diverse as machine tools and construction equipment revived with the arrival of the multinationals.

Incoming engineers: Nissan's UK manufacturing plant at Sunderland

Not all of the inward investment was in the form of new arrivals. Foreign ownership of established UK companies became commonplace. Sometimes, especially during the recession, British firms were picked up cheaply by their rivals; sometimes takeovers were part of a grander strategy to create Europe-wide corporations. The buying and selling of companies and the transfer of assets between Britain and the rest of the world gathered pace in the recession, but it was a phenomenon which was here to stay.

At times, some engineers worried that large parts of British industry seemed to be passing into foreign hands. There was concern that the new owners would be less committed to Britain and to British manufacture than previous managements had been. It was taking a while for the realisation to filter through that the engineering industry worldwide had changed for ever, and that competitiveness in product and performance was the best guarantee for continued survival.

Incoming technology: British Leyland linked with Honda of Japan to produce the Triumph Acclaim

The traffic in company buying and selling was in any case not all one way. British companies keen to take their place in global markets – and in many cases keen too to reduce their dependence on a domestic British market which was of limited size and uncertain stability – pushed outwards into continental Europe and beyond, often by acquisition. By the mid 1980s, British companies were the single biggest foreign buyers and investors in North America, eclipsing the Japanese.

Inward and outward investments were important parts of the wider internationalisation of engineering as an industry and as a profession, and the institutions had an important part to play in that. Though there had long been contacts with overseas professional bodies, and speakers at Institution events had been drawn from many countries, it was only in the early 1960s that a president of the IMechE made the first official overseas visit, and the countries visited then were largely Commonwealth nations linked to Britain and the IMechE by an imperial past rather than a technological and industrial present or future.

But within 25 years of that first visit, it would be inconceivable that the newest and youngest member of the Institution, let alone the president, should not be aware of worldwide trends in engineering technology and in the wider business of engineering industry.

Britain's membership of the European Union was a big factor in widening the horizons of engineers and engineering industry: by the end of the 1980s, more than half Britain's trade was with the other countries of the EU. Europe had an impact in professional matters too as engineering qualifications gained in one country were recognised in others. Some members of the IMechE, noting that professional engineers appeared to be accorded more respect in other

A two-way process: British companies have been consistently successful in overseas contracts in many engineering sectors. This is the new Tsing Ma bridge in Hong Kong

European countries, started using the international "Eur Ing" designation through FEANI, the European association of national engineering bodies, to indicate their professional status.

The European ideal was more than just symbolic: in the mid-1980s, work started yet again on a scheme to link Britain to mainland Europe by a tunnel, and this time, though the scheme went badly over budget and the connecting infrastructure work on the English side was far from ready, the job first talked about in the Institution more than a century earlier was completed. The Channel Tunnel

European partners: civil air-liner builder Airbus Industrie was part of a wider move to the internationalisation of industry

opened for through trains in 1995; Britain was physically connected to the rest of Europe and engineers of many disciplines had been the means by which the connection was made.

Back in Britain in the 1980s, international inward investment was an important part of the slow economic recovery from recession. It also contributed to a slowing-down, if not yet a reversal, of the long decline of manufacturing which dated back virtually to the Industrial Revolution. And it was a significant factor in the overall cultural change in engineering.

But alongside the growing number of global engineering companies in Britain there were other changes in corporate culture. Since Victorian times, most engineers and most engineering industry workers had been employed in big companies: this trend started to

Submarine entente: Britain was finally linked to mainland Europe by the Channel Tunnel, opened to traffic in 1995

A new kind of train: Continental-style Eurostar trains connect Britain with France and Belgium

reverse. There was a rising generation of new small companies able, often through the use of computer-based technologies, to play disproportionately important parts in big projects.

Partly as a reaction against the dire troubles of the massive monolithic firms of the 1970s, many giant companies also began transforming themselves into collections of smaller customer-focused businesses. Others reorganised themselves internationally, by going into alliances with former competitors and tackling Europe, rather than just Britain, as the home market.

The increasing numbers of smaller companies in engineering industry, and the dismantling of the giant corporations with their centralised training departments, have put a greater burden on bodies like the engineering institutions to provide an effective service in information and continuing education to individual engineers.

The IMechE's continuing professional development programme in fact predated the 1980s, but in the early years there seemed to be almost as many courses cancelled for lack of interest as there were successes. The numbers looking to the Institution for this kind of service were limited, partly because the IMechE had not really been in this business before and partly because it took the sharp shock of recession to make many engineers aware of the need to update their technical knowledge: but the Institution persisted, and the market emerged.

What also emerged, though, from the continuous review of Institution activities across the 1980s was that the IMechE was not necessarily the best body to be running courses. Its role was to set the standards and accredit courses provided by educational establishments and training companies whose main business was professional training. Conventional courses, though, were only part of the moves into continuing professional development: the ever-expanding conference programme was also linked into structures put together by the engineering institutions and the Engineering Council to ensure that engineers' knowledge was updated systematically.

But it was not just in terms of technology that engineers' knowledge required constant refreshing. The economic shock waves of the oil price upheavals of the 1970s, followed by the two recessions of the early 1980s and the early 1990s, put a new emphasis throughout industry on finances and required from engineers extra skills in management, cost awareness and teamwork.

This was a point which had been made at the end of the 1970s in the Finniston report and which was, for many engineers, one of the less palatable conclusions of that report. Finniston's view was that engineers had become too divorced from general business processes: much of this was to do with the lack of appreciation of engineering by non-engineers, especially those in management roles. But engineers themselves could not be entirely absolved from responsibility.

In the new climate of economy, the apparently unbridled pursuit of technology, which characterised the 1960s, had to be reined back, and the engineering developments and innovations which have taken centre stage since the start of the 1980s have largely been those associated with manufacturing efficiency, design, process control and environmental concerns. And in this, engineers have had to be as aware of customer demands and market conditions as the other functions within a management team: they have had to be accountable.

Alternative energy: commercial windpower, as in this Cornish wind farm, made environmental sense and, increasingly, economic sense

The move towards accountability has taken many forms, and not all of them have been financial. The responsibility of engineers for the products they develop, for the processes needed to manufacture them, for the workplaces where they are manufactured and for restraint in the use of raw materials and energy sources has been enshrined in legislation ranging from environmental laws, through health and safety regulation to product liability directives.

Energy alternative: the Sizewell B nuclear power station in Suffolk. Nuclear remained the biggest alternative to fossil fuel fired stations

Back in the 1970s, the IMechE had been quick to seize the initiative in one area, setting up the Pressure Vessels Quality Assurance Board in response to demands from the nuclear power industry for someone to take responsibility for developing standards in an important part of mechanical engineering. As with CPD, the Institution's role was as a catalyst, rather than as a long-term participant, and the board's successor body is no longer part of the IMechE. The 1980s provided new activities that accountable engineers were responsible for, and the IMechE had to respond.

At times, this trend towards economy, efficiency and accountability has seemed to inject an unwelcome utilitarian austerity into the engineering profession, and has appeared to militate against expressions of flair and imagination. And in some cases, the demands of economy directly affected engineering projects which involved innovative technology.

The Advanced Passenger Train was an example. It was an ambitious attempt to advance UK railway technology on several fronts in one go, but it was possibly too ambitious for the financiers, and patience and money both ran out before the project had managed to prove itself. It was a high profile setback for high technology: in fact, though, neither the project itself nor its probably temporary failure were typical.

Running away on right lines? The Advanced Passenger Train had several high profile failures. Yet some of the technologies it pioneered still hold promise

The new engineering of the 1980s was less to do with great leaps forward through invention and much more to do with the measured and incremental approach to innovation geared to maximising technology benefits. In many respects, this change was good news for the Institution of Mechanical Engineers.

The Institution has a long record of running lectures and conferences on specific technologies, and certainly within some of the more product-oriented groups, particularly in recent years in the aerospace and automotive areas, has not been afraid to tackle quite specialised topics at the leading edge of technology.

Throughout its existence, though, the Institution has had to balance the requirement to push the boundaries of engineering technology – to help those engineers working at the leading edge – with the needs of the much greater number of mechanical engineers whose work, mostly in engineering industry, is some distance back from the frontiers of science and technology.

This experience of dealing with the broader issues that affect mechanical engineers working across the range of industrial sectors was particularly relevant in rebuilding British engineering after the ravages of the recession of the early 1980s. Subject areas such as design, manufacture, control – issues that involve management and the integration of disciplines – had long been well represented in Institution events. And one consequence of the increasing internationalisation of engineering industry and the need for British engineering to address the issues of international competitiveness was an increasing demand for information on areas such as supply chain management and quality assurance where overseas companies appear to have set the pace.

The universal demand for information on "management" topics like these led in 1988 to what might have appeared to be a strangely perverse decision. The Engineering Management Division set up in the restructuring of the early 1980s was scrapped. The reasoning was tactical – attendance at meetings had been disappointing – but it also fitted with the wider strategic view. This was that engineering management was not a topic, like aerospace, the automotive industry or railways, which could be confined to a section of the Institution's membership by being discussed only within the constraints of a single division. It was a subject which applied universally, to all areas of the Institution. Mechanical engineering in the late 20th century, the IMechE was saying, was about engineering management, as well as the more specific industries and engineering technologies.

Many of the engineering management ideas that the Institution has introduced to its members in recent years have come through the internationalisation of industry. The process has, of course, always been two-way: the pioneers of the Institution were by 1847 already involved in international contracts and overseas engineering development, and that tradition has been maintained. And there is nothing new about Britain importing engineering management techniques from other countries: the Americans, for example, introduced mass production to Britain through the motor industry and many of the tools of productivity measurement and piecework originated in the scientific management ideas of Frederick Winslow Taylor at the Bethlehem Steel Company in turn-of-the-century Pennsylvania.

But it has not been the Americans who have been responsible for the big changes in engineering management in the past 20 years: it has been the Japanese influence that has been credited with leading the change from the mid 1970s onwards.

When Japanese versions of consumer goods and cars started appearing in Britain in the 1960s, they were widely derided as inferior imitations. The story of how Austin, the British car manufacturer, had sold the designs for the A40 Somerset – scarcely a world-beater of a car – in the early 1950s to Nissan when the Japanese company was looking for a new model to move back into volume production was well-known, and was taken in some quarters as evidence that the Japanese were no threat, either as designers or as manufacturers.

Well before 1980, the view of Japanese goods and Japanese industry had swung through 180 degrees. On quality, reliability and price, Japanese manufacturers were more than matching UK and European companies, and in the ability to take new technologies and turn them into marketable products, they were setting the pace. More than that, the whole of the hugely successful Japanese economy was based on a fast-growing manufacturing industry.

Even when the change in Japan and its goods was recognised,

though, the full implications were not immediately digested, and various "reasons" were put forward for the country's sudden success. Not all of the reasons stood up to close examination.

Excuses about low cost bases, which could be levelled at other Far East economies, did not apply to Japan. The Americans in particular became bothered by Japan's continued reluctance to open up its domestic markets to goods of foreign manufacture, arguing that a protected domestic economy gave Japanese companies the solid foundations from which to build colossal export-oriented manufacturing businesses. There was some justice in that argument, it did not answer why it was that Japanese goods managed to be so competitive in export markets worldwide. There was also no substance in unflattering references to Japanese working practices and conditions.

To a degree, Japan's success as an economy founded on manufacturing industry was unsettling in Britain and in other Western countries partly because it appeared to have been achieved so quickly. But it was also unsettling because it flew in the face of a growing conviction that the mature economies of the West – a list which naturally included Britain, the first nation to go through the industrial revolution two centuries earlier – had reached a "post-industrial" state where national economic vitality and viability was to be found in activities other than manufacturing.

Chancellor Nigel Lawson was not alone in seeing manufacturing as a wealth-creating option, not an essential; others had openly wondered since the 1960s whether manufacturing was a phase that economies passed through on their way to "maturity".

But now if Japan, by no stretch of the imagination a developing country, was building its economy on manufacturing, was it right to see manufacturing as a dispensable part of our economy? And if Japan could transform itself through manufacturing excellence into a world economic power in little over a generation, could we not do likewise? From the mid 1970s onwards, government ministers, top engineers, industrialists from companies big and small, and economic commentators began regularly filling aircraft on the Tokyo routes, all hoping to discover some of the secrets of Japan's economic miracle.

Early reports back from Japanese industry on how the manufacturing success had been achieved were often confused. Concepts such as kanban and just-in-time manufacture sounded initially complex in terms of organisation – and then deceptively simple in practice. The fact that one of the great mysterious secrets of Japanese industry was a simple docket attached to a component which charted its progress through the manufacturing process seemed almost ludicrously obvious. The idea that a factory should manufacture only what the market wanted at the time when the market wanted it was scarcely a revelation.

Yet if the individual components of the message from the Far East seemed self-evident when they were expounded in meetings at the engineering institutions or in articles in the technical press, the cumulative effect was significant. With economic pressures already forcing engineering companies to seek ways of getting more from less, the holistic Japanese approach, looking at the engineering company as a single entity and not as a collection of discrete operations, had logic as well as a successful track record.

The logic was backed up, too, by the convergence of engineering disciplines under computer control. Computers, or some form of proto-computer automation, had been used in measurement and control systems since the late 1940s; numerical control systems for machine tools, using punched cards or tape, had been developed across the 1950s; robotics, or computer-controlled handling and movement, reached a basic level of viable operation by the end of the 1960s; computer-aided design had started in rudimentary form by this time as well; and computer systems were being used for office and business routines such as payroll, personnel and paperwork.

Shopfloor revolution I: Rolls-Royce fitters use information from a computer-aided design screen

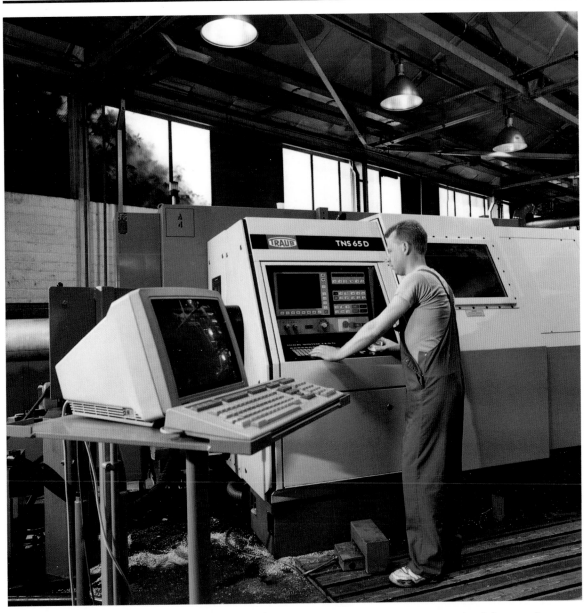

Shopfloor revolution II: computer systems such as this one at British Aerospace link management control systems with work on industry's shopfloors

In the early stages, computer systems were prized for two features above all others: quick calculation and storage capacity. Many of the early computing functions merely accelerated existing operations and systems, though the speed of calculation also allowed greater complexity to be built in. The ability to store data meant that tasks could be repeated accurately, and also provided reliable archives.

But often, these early computer-based systems were asked to do little more than the human-based operations that they replaced, even if they might do the same tasks more quickly and more accurately. Early CAD systems, for example, were basically a drawing board on a screen, and used in much the same way as a conventional draughtsman's board.

As with so many innovations of the past 25 years, it is all but impossible to put a precise date on the point where computer systems used for discrete parts of the engineering process were put together within a single computer system, allowing the development of a new phase

Shopfloor revolution III: look, no hands. Robot use in industry was not the stuff of science fiction. But the precision of robotic handling has been a factor in raising shopfloor quality and cutting shopfloor staffing levels

of computer-based operations in which design, manufacture, process control and now office functions such as sales and marketing are integrated.

The first tentative steps were made in the 1970s, and CADCAM – computer-aided design and computer-aided manufacturing – was a commonly used term long before the final links were made which would allow CAD-derived data to drive manufacturing systems. Certainly an important factor was the explosive growth in available computing power which the microprocessor allowed, with costs plummeting as capacity soared. Hand-held devices now have greater capacity than machines which occupied whole rooms did 30 years ago. In the 1990s, the trends show no signs of slackening.

In parallel with the computing revolution came the communications revolution, which helped to overcome the constraints of geography. Global companies can now act globally and specialists can be called in for specific tasks no matter where they are located.

Increasingly sophisticated computer-aided design is used on all manner of products

The computer revolution in engineering industry has been in terms not just of what is done, but also of how it is done. In parallel with the engineering management thinking derived from Japan's manufacturing success, present-day computer systems encourage the broad "holistic" view of engineering. The Institution of Mechanical Engineers has contributed materially to this process: it is no accident that its long-running series of CADCAM conferences goes under the title "Effective CADCAM", indicating that the important thing is not the technology itself, but what it can do.

In some respects, of course, the new holistic approach in which engineering functions are part of a business-wide computer-based integrated system has blurred some the distinctions, both within the engineering profession and with other professions.

This blurring of distinctions is on several levels. Mechanical engineers have had to understand the abilities of electronic systems to build them into their own machines; they have also been able to use electronic or computer-based machinery to improve the control of mechanical engineering processes; and they have been called upon to develop new technologies themselves as yet smaller electronic devices, down to atomic size, are developed, requiring machines on a wholly new nanotechnological scale for their manufacture.

On yet another level, computer systems and a holistic management approach have developed ideas such as concurrent engineering, where the various stages of research, development, design, process planning and manufacturing are carried out virtually simultaneously: the technique was first demonstrated in Britain on Ford's Zeta engine project in the 1980s and is now being used on highly complex engineering projects such as the new generation of nuclear submarines.

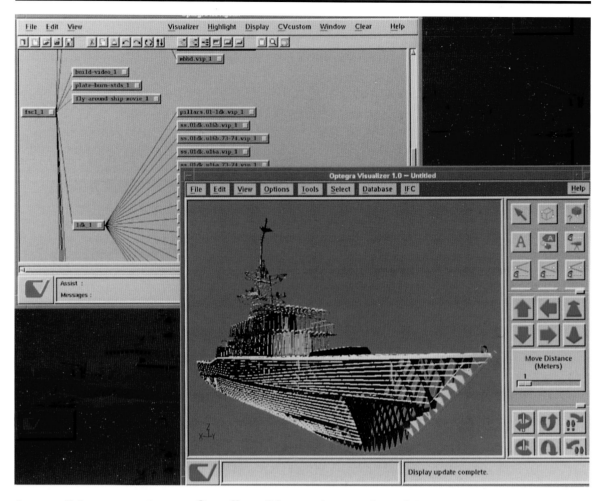

Computer links: systems such as Computervision's Electronic Product Development combine computer-aided design with parts information, manufacturing schedule and management functions on projects as big as naval ships

One effect of the coming together of the engineering disciplines on the IMechE is that a very high proportion of conferences and other events are now held in conjunction with other engineering institutions, or with bodies representing other professions. Some of the specialist subject groups or divisions are formally linked with similar groups in other bodies, both inside and outside the engineering profession.

It was, of course, one of the original tasks of the Engineering Council, as envisaged in the Finniston report and preserved in the rather watered-down version of the Finniston authority which was eventually enacted by the government, to attempt to reduce the multiplicity of engineering institutions. Progress on this, despite the changes in technology and in working practices which have brought different parts of the engineering profession closer together, has been slow. From an original total in the mid 50s, the number of institutions has dropped by only around one a year to 40.

Conferences and joint activities illustrate, though, that closer working with other engineering institutions does not have to mean amalgamation or takeover. The IMechE and the Institution of Civil Engineers have had some formal and some informal agreements over the sharing of some facilities such as catering for a number of years. Separately, there have been occasional arrangements with other engineering institutions: the Institution of Electrical Engineers, for instance, was the first of the institutions to make the great leap forward into computer ownership, and the IMechE "borrowed" disk space for the first ever computerised list of its members.

Perhaps more surprisingly in the 1980s, when there appeared to be real momentum behind the idea of institutional mergers, discussions with the Institution of Production Engineers, which had renamed itself the Institution of Manufacturing Engineers, came to nothing. Some IMechE members were unenthusiastic about a name-change for their own Institution.

Progress at a rate of knots: an example of concurrent engineering

In recent years, other potential liaisons and collaborations have been mooted. In the early 1990s, there seemed for a while a real chance that discussions with the Institution of Electrical Engineers might lead to a merger which would create a single institution for the vast majority of professional engineers working in industry. The discussions foundered again, though the IMechE council voted in favour of the merger. But they were instrumental in opening the door for Sir John Fairclough to embark on the restructuring of the Engineering Council. And more recently a memorandum of understanding with the IEE has reopened the way towards closer collaboration.

The arguments for and against institutional mergers are complex, and seem consistently to have generated more heat than light. Converging technologies have made links with other engineering institutions more attractive though conversely, with so much collaboration already between the different bodies, the pressure for formal amalgamation may have lessened.

Indeed, it may be that the new Engineering Council, recast in 1996 very much as the umbrella body for all engineers and all institutions and with the intention of furthering collaboration without endangering diversity, will make further rationalisation of the institutions less of an issue. In addition, the Royal Academy of Engineering continues to provide a focal point for distinguished engineers from all branches of the profession, and has in recent years done influential work on issues which cross the boundaries within the profession and beyond it.

The continuing complaint which has echoed down the years is that engineers – not just mechanical engineers, but all engineers – have not had the influence in government and on the public perception that they deserve. The suggestion has been that bringing the different branches of the engineering profession together might increase this influence and improve the perception. Until and unless it happens, there is no way of telling whether the suggestion has merit or not.

But in the mean time all engineers have the satisfaction of knowing, and being able to demonstrate, that their talents, more than those of any other profession, have shaped the world we live in and contributed to the progress of mankind.

8 Progress into the Future

1997 AND BEYOND

At the end of its first 150 years and on the brink of a new millennium, both the Institution of Mechanical Engineers and mechanical engineering itself have come a very long way from the time that a small group gathered at the Queen's Hotel, Birmingham, to share knowledge of mechanical science and "give an impulse to Inventions likely to be useful to the World".

The progress of the IMechE itself has been beyond anything that the founding members could have imagined. Membership in the 150th year stands at more than 78,000; the programme of meetings, which started with just four gatherings a year, now extends to many hundreds of events in many varieties of format; engineers across the world can find out what is happening in their Institution in an instant simply by dialling up the IMechE's World Wide Web site through their computer terminals.

But if the progress of the Institution has been astounding, then just as impressive is the fact that the aims of the IMechE today differ little from those in the advertisement that brought the original members to

Plus ça change... London Bridge in the 1860s

the rush hour
It was ever thus. Old London Bridge, Monday morning, April 1867.

...plus c'est la même chose: the motorway on a sunny summer day in 1993

Birmingham in January 1847. The sharing of knowledge about mechanical engineering and the promotion of innovation are the business of the Institution of Mechanical Engineers and its members today, just as they were 150 years ago.

It is tempting to think that, were George Stephenson to return in 1997 to the Institution he helped to found, he would find the organisation and the principles that underpin it familiar, to a degree at least. He would recognise the basic mechanisms of discussion and debate, the fundamentals of sharing information and knowledge for the common good, and the commitment to engineering excellence and innovation. He would marvel at the sheer scale of the operation and at the range of modern IMechE's activities, and he would be nonplussed, though doubtless excited, by some of the methods used to provide services to members.

But in its aims and objectives, and as an organisation composed of the top people of the mechanical engineering profession, the IMechE of 1997 is recognisably the same as the IMechE of 1847.

Where Stephenson would be totally lost, though, would be in the mechanical engineering of today. As a subject, he would find it bafflingly complex, founded on scientific principles many of which were unknown in his day, using materials and technologies that he would not recognise to achieve feats of strength, endurance, stamina and even intellect that he would not have dreamed possible.

When he wrote the official *History of the Institution* of Mechanical Engineers for the IMechE centenary in 1947, R H Parsons, a member of one of the most famous mechanical engineering families of them all, attempted to describe the state of mechanical engineering in 1847. He found it simpler to identify the materials and the technologies that the mechanical engineers of 1847 did not have, rather than those that they actually used.

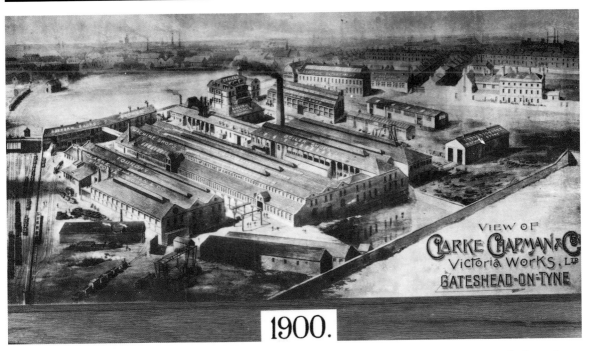

Factory progress: within the IMechE's first 50 years, manufacturing moved from being an industry of small workshops to huge factory complexes, such as Clarke Chapman on the Tyne

What today might be considered the basic raw material of mechanical engineering, mild steel, was, for instance, invented almost a decade after the IMechE was formed. Other metals now regarded as commonplace, such as aluminium, were laboratory curiosities. Mineral oils for fuel were first extracted in 1848, and there was no drilling for petroleum for a further 10 years. Electricity was used for telegraphic messages, but for nothing else, and electricity generation was by means of batteries or crude magnetos. Sir Joseph Whitworth, president of the IMechE for three years from 1856-57 and in 1866, had started the process which would lead to standardisation of fastener threads and machine tools, but his wider ideas on the necessity for accuracy and precision in measurement standards did not reach a wide public until the Great Exhibition of 1851. Among machining techniques, grinding and milling machines were among those not yet available.

Even in basic science, knowledge in 1847 was limited. Parsons points out that at this stage in the history of mechanical engineering, heat was widely believed to be a substance, called "caloric", to be found in the pores of bodies that were hot. The results of Joule's experiments which led to a mechanical equivalent for heat were published in the same year as the IMechE was founded, as were Regnault's first steam tables.

The wonder of the founding fathers of the Institution of Mechanical Engineers is that they were able to achieve as much as they did with the limited range of materials and the often crude techniques that were available to them. In particular, the ability to make large-scale stationary steam engines and workable locomotives – which were, of course, subject to the different stresses and strains of locomotion and traction – from the materials and machines then available was quite remarkable.

George Stephenson's centenary in 1881 was celebrated with a parade of his locomotives at Newcastle upon Tyne. The locomotive second from the right worked at Hetton colliery and is probably the one pictured in Hare's engraving, reproduced in Chapter One

Nevertheless, the range of mechanical engineering in 1847 was quite restricted. The steam engine was the paramount machine of the age, and was applied to many tasks including – thanks to George Stephenson – transport. Running off the back of this universal power source, or using other sources of energy including animals and human beings, there were machines for industries such as textile manufacture and for some agricultural activities such as threshing, and the beginnings of machine tools for large-scale metal goods

John Ramsbottom, IMechE founder member, one of the first to present a paper on an invention, president – and long survivor

manufacturing to go along with the precision tools which were inherited from craft activities in watch and clock-making and scientific instrument manufacturing.

In 1847, mechanisation was being applied to transport on the railways and increasingly in ships. It was used in some parts of manufacturing, though mechanisation was only one of a number of reasons why particular industrial sectors adopted the factory system during the Industrial Revolution. But mechanisation had penetrated the mid-Victorian home hardly at all and had made little impact on a whole range of human activities such as construction and communication; agriculture, often overlooked as the single biggest "industry" of them all, remained for many years an odd mixture of some activities where there had been early mechanisation and others where there was, until well into the 20th century, virtually no mechanisation at all.

If the wonder of the mechanical engineers of 1847 was that they achieved so much from so little, the wonder of the succeeding 150 years has been that mechanical engineering has made so much progress into so many different areas of human activity. Particularly impressive was the degree to which it expanded its repertoire and range during the lifetimes of those engineers who formed the Institution.

John Ramsbottom is a name that recurs several times in the pages of the Institution's history. He was the London and North Western Railway's chief locomotive engineer and the originator of the six-coupled goods engine, safety valves, piston rings – and of the water troughs in the middle of railway tracks used for high-speed water pick-up by express locomotives. Ramsbottom was one of the original members of the IMechE in Birmingham in January 1847, and like George Stephenson was an early example of the self-made engineer of little formal education but immense practicality. His credits include the delivery of the first paper describing a true invention to the Institution, in 1854, and he followed this with several more in the next few years. He became president of the Institution in 1870/71, and died at the age of 82 in the year of its golden jubilee.

Ramsbottom's lifespan, though not exceptionally long, encompassed an enormous range of mechanical engineering inventions, and the mechanisation of huge areas of human activity and endeavour. Within Ramsbottom's lifetime, mechanical engineering progressed, almost literally, from the blacksmith's forge to the modern factory, from the hammer and anvil to the precision machine tool. And the Institution of Mechanical Engineers, formed to "give an impulse to Inventions likely to be useful to the World", was one of the main forums in which the great engineering inventions of the Victorian period were revealed and discussed.

By the time of Ramsbottom's death in 1897, it is quite likely that he would have seen, and possibly even ridden in, one of the earliest motor cars – vehicles composed largely of a material not available in 1847, powered by an engine using a technology not thought of in 1847 and employing a fuel not used as such in 1847. In the year of his death, the first steam turbine powered ship, the Turbinia, was launched. Ramsbottom may personally have just missed domestic developments such as the electric light and the telephone, both of which had been invented by the time of his death, but which were not in widespread use in 1897. He certainly missed, though only by a few years, the first powered flight.

The sheer inventiveness of the mechanical engineers of Ramsbottom's generation, however, has left their successors with a problem. It's a problem that affects engineers from other disciplines too.

In popular perception, the great days of engineering were in the nineteenth century. The great engineers of popular culture – the few engineers whose names are widely known – are nearly all mechanical engineers. But most of them, and engineers like Ramsbottom, who were famous in their time, have been dead for 100 years and more. And today's mechanical engineers appear, in comparison, to be less inventive and therefore less worthy of renown.

The idea that this century's mechanical engineers have somehow been less innovative than the famous Victorian engineers ignores the fact that there is a pattern that all new fundamental technologies follow in their development. The pattern is that the opening up of a new branch of technology is followed by a period where that new technology is applied to fundamental human activities and basic problems, and the result is a flurry of inventions.

Mechanical engineering as a discipline and a technology in its own right, based on its own discrete section of science and combining scientific principle with artistic flair and practical craft, was primarily a Victorian development, and its growth was, of course, helped in great measure by the establishment of the Institution of Mechanical Engineers.

It is even tempting to suggest that before the IMechE there was no such subject as mechanical engineering: tempting, but not strictly true, since the idea of machines that enhance human capabilities goes back to the earliest levers and winches. It would be true, though, to say that 1847 was the beginning of mechanical engineering as a separate discipline, as it was the year in which not only was the Institution founded, but also the first professorial chair in mechanical engineering was set up at University College, London and occupied by Eaton Hodgkinson, one of the IMechE's first honorary members.

So the great Victorian names of engineering – the Stephensons, Whitworth, Parsons, Armstrong, and Ramsbottom – were pioneers of what was then a relatively new discipline. Part of their pioneering was to apply the new discipline to basic problems of power, materials, transport and communication. And, as with any new and fundamental technology, the efforts of the Victorian mechanical engineering pioneers were rewarded with a host of important inventions.

There is a parallel in the second half of the 20th century which gives the lie to any suggestion that engineering inventiveness has somehow disappeared since Victorian times.

The cascade of applications and inventions which have followed on from the development of the computer is analogous to the explosive growth in mechanical engineering inventions a century earlier. In fewer than 50 years, computer systems have penetrated into a vast array of human activities and industries. They have been used in design, manufacture, measurement and control systems, even art. Their impact has been accelerated by greater sophistication in communication than the mechanical engineers of 100 years ago could call upon, and to which the computers themselves have contributed.

One difference between the surge of mechanical engineering inventions of the 19th century and the more recent computer-based inventions has been that where the Victorians attached names of individuals to inventions, we tend to attach names of companies or product brand-names. Neither practice is particularly fair to the large majority of unsung engineers: the great Victorian mechanical engineers relied on their engineering teams to convert their ideas into practicalities; the corporate R&D teams of today rely on often-unnamed individuals for the creative spark.

But the analogy between the development of mechanical engineering from the mid-19th century onwards and that of computing from the mid-20th century onwards is a close one. Both developed from a scientific base, but required a new way of thinking or of ordering the known science. Both were given impetus at a relatively early stage in their development by the discovery of a new material which facilitated widespread adoption – steel in the case of engineering, silicon for computing. Both went through a remarkably fertile stage of inventiveness.

The actual inventions, though, are only the start of a process, usually extending over many years and involving many more people than the original invention, in which an idea is refined and developed and any products that result from it pass into widespread use.

In many cases, the actual moment of invention is hard to pin down, or goes by wholly unnoticed. In others, it is disputed. Americans claim that the incandescent electric light bulb was an American invention by Thomas Alva Edison; the British view is that Edison was beaten by Joseph Swan.

But perhaps as important as determining the actual moment of invention is that, over the next 50 years after Swan or Edison made the first light bulb, many thousands of jobs were created in a new worldwide manufacturing industry, and domestic life changed for millions. Engineering processes allowed teardrop-shaped globes of glass and thin wire filaments to be brought together in mass production and also organised the business of construction so that, within a few years of the invention, each room in each new building came equipped with an electric lighting socket. These processes did not bring fame to individual engineers, but they were no less vital than the initial spark which gave immortality to Edison and/or Swan.

The inventive stage, then, is part of a longer process of technological development, or innovation. It is the stage from which the whole process begins, of course, and is important in its own right as a means of gaining competitive advantage for companies and individuals – though often the pioneers lose out in the exploitation of an invention because others can take advantage of the mistakes inherent in any trial-and-error method of development.

But it is through this longer process of innovation as a whole that mechanical engineering makes its great contribution to progress.

Innovation is not an easy word to define. Much of the work in innovation that the Institution of Mechanical Engineers has been involved in over the years has had to do with process and product improvements rather than new products or engineering inventions. It has involved techniques such as "productionising" – an ugly word denoting the bringing of a product from prototype or one-off to mass or

Present-day innovation: Rolls-Royce's Trent aero-engine family

series manufacture – and process optimisation, and it has meant the kind of lateral thinking that can apply devices or materials from one sector in wholly new areas. It has to do with components and sub-assemblies as much as with finished and recognisable products and it is about perspiration as well as inspiration.

There is a long public tradition in Britain, which engineers have not been able until relatively recently to make much impact on, of ignoring the wider definition of innovation and equating it with inventiveness. The confusion has led to an long-standing criticism of British engineering: that British inventiveness has not been matched by an ability to make money from new technologies or products. Given relative industrial performance against other countries, there may be some justice in the stricture, but it probably has as much to do with the status of engineers in industry and business as whole.

The IMechE has in any case rarely been in the business of helping in the development of specific products or inventions. Even in Victorian times, when great names in engineering invention were regular attenders at Institution meetings, its role was as a channel for the dissemination of information about new ideas. Many of those ideas are about taking inventions to production, or about refining products and processes. So the Institution's role has always been strongest in the improvement and development phases of the innovation process, rather than at the point of invention.

This strength has had to come even more to the fore in recent years as the emphasis in industry has switched from technology for its own sake to increasing efficiency and competitiveness. But it can be seen from the earliest days in the spread of mechanisation and of mechanical engineering products into new areas. If inventiveness appears to have been the hallmark of the Institution's first 50 years, the whole of the first 150 years have seen the Institution working to spread the influence of mechanical engineering much wider and helping mechanisation to permeate through almost every sphere of human activity.

The Institution has been an important part of the process of bringing engineering knowledge and information into the core applications of design, manufacture, measurement and control. It has also been central to the process of broadening the remit of mechanical engineering so that mechanical knowledge and skills are applied to products and systems developed in other branches of engineering and beyond: in areas such as medicine, for instance. And, conversely, it has frequently been the channel for receiving ideas from other parts of engineering and technology which can be applied to mechanical engineering systems and artefacts.

All of this is rather a long way from the simplistic perception that engineering is about invention, and that engineering success or failure – individually, corporately or even nationally – can somehow be measured merely by counting inventions. It is a much bigger subject, and a much more complicated and unpredictable one.

A lot of the fascination of engineering that engineers often seem to find difficult to explain to non-engineers is that, behind the rigorous application of engineering logic to problems and the scientific method of research and development, there is an engaging unpredictability about the whole process. Exactly why particular inventions happen at particular times is far from clear; why some ideas take years, even centuries, to bring to fruition when others seem to race from prototype to mass production is a subject for academic treatise. Innovation is not a straightforward straight-line graph of a process, and the degree and direction of the progress that comes from engineering innovation is not easy to forecast.

Technology forecasting of all sorts is an imprecise science, and invention is itself an uncertain business. There is a saying that the person who invents a better mousetrap finds that world beats a path to his door. Many of the engineers who have, over the years, invented or assisted in the development of new products or technologies can confirm that this is by no means a universal truth, and the history books are littered with products for which immense and intense demand was predicted, but which flopped, from the Sinclair C5 electric vehicle to moving pavements for urban streets.

Part of the role of promoting engineering innovation has always involved the Institution in the risky task of picking promising technologies – indeed, its everyday business of organising conferences and other events is about selecting topics that may turn into winners. The Institution's overall record of picking technologies with potential to develop is pretty impressive.

But inevitably there have been times when the Institution has helped to pursue ideas which turned out to have no future. And at other times it has been part of a wider process of over-estimating the importance of a particular development or of discussing issues using arguments which were based on false premises.

Records of past technology forecasts are, in any case, a rich source of amusement, and sometimes embarrassment. The simple fact is that the future direction of engineering technology has proved, over the years, no easier to predict than any other sphere of human behaviour or activity. Engineers may have a better record of forecasting technology than science fiction writers – and many of the better science fiction writers are, of course, engineers – because they have a better appreciation of technical issues and realities. But visions of the future are rarely about technology alone.

An example of how a false assumption in one area unrelated to technology could lead to some arrestingly incorrect technology forecasts came in one of the few events over the years in which the Institution of Mechanical Engineers has gone in for long-range projections: an Automobile Division discussion in 1965 on "Transport in AD 2000".

Many of the discussion's conclusions now, in fact, seem remarkably sensible. It decided, for instance, that there would not be dramatic change in the relative importance of different transport forms and that the internal combustion engine would continue to be the main driving force for transport systems.

However, one of the base assumptions behind the discussion was that the population of Britain would rise from the 53.5 million of the mid-1960s to around 75 million in 2000: a reasonable assumption given the increase in population in the 20 years from the end of the Second World War. Transport policy and transport systems, therefore, would have to be geared to a big increase in demand and would probably, if life in the south-east of England was not to become unbearable because of congestion, require new cities to be built away from the London sprawl as the new towns in England, Wales and Scotland had been built in the years leading up to 1965.

The reality, of course, is that population growth has been rather less than a quarter the expected level. And though Milton Keynes has been developed from greenfield site to new city in the period since the IMechE meeting, it has not had to be replicated elsewhere.

Because this basic assumption about life in Britain in the year 2000 proved false, some of the other predictions from the 1965 discussion have fallen by the wayside also. Some, in any case, broke a basic rule of technology forecasting by being too specific. So the speaker who forecast that traffic congestion would only be solved by "bringing in the third dimension" – cars that could fly, protected from aerial collision by electronics – might be embarrassed to read his prediction today, though at the time there was undeniable logic in the suggestion if the forecast population growth had come about. The same speaker might be more reasonably embarrassed by a second suggestion he made, that nuclear supersonic airliners would be flying by the year 2000; in fact, US attempts to build nuclear-powered aircraft had already been abandoned by the time he spoke, because the weight of the shielding precluded take-off.

A different potential problem for bodies like the engineering Institutions which aim to be one step ahead of the trends in technology is the strong possibility of exaggeration. Technology is as subject to prevailing "fashions" as any other activities, and the importance of individual engineering devices or of whole technologies can be miscalculated.

In Britain in the 1950s, for example, there was much discussion about future transport systems, as there was a decade later. The railways were conscious that they were nearing the end of the steam age; diesel traction was unproven and electrification was judged to be expensive. Studies were commissioned to determine whether other forms of transport were likely in coming years to take some of the railways' traditional passenger and goods markets away, thus changing investment equations.

Most of these concluded, sensibly, that road transport was a threat particularly on short journeys and for goods transport; more surprising to modern thinking was a second conclusion, repeated in several well-argued reports, that long-distance passenger rail routes would face intense competition from "personal helicopters".

The helicopter was the "fashionable" transport technology of the years immediately after the Second World War. It was expected that each town would have its heliport, and that civilian transport applications would outweigh by far any military usefulness. In fact, the opposite was true: the civilian potential of the helicopter was overstated, but the military importance increased significantly as warfare changed from conventional battlefields to guerrilla operations.

The helicopter brought the joy of being able to fly while keeping one's hat firmly on one's head: an early Sikorsky, used as illustration for the helicopter pioneer's 1955 IMechE lecture

False forecast: by the mid 1960s, a heliport in every town seemed a far from rash prediction. London's was, and still is, at Battersea, on the opposite bank of the Thames from the Lots Road power station

Like the helicopter, ultrasonics, using ultra-high frequency sound, was a technology which fleetingly appeared have much wider application than turned out to be the case. In the mid to late 1960s, it was frequently mentioned alongside electronics as a fundamental technology building block for future systems. There were also going to be domestic applications: there were confident predictions that ultrasonics would bring dry-cleaning into every home, and wilder forecasts which suggested that personal hygiene would benefit from the replacement of the shower unit with a dry ultrasonic version in which the dirt would be literally shaken off by a bombardment of ultrasonic waves. In the event, ultrasonics found solid niche applications in medical imaging, defence and component cleaning, but the other ideas have yet to be fulfilled.

The art of picking winners in technologies is in recognising when there is a fundamental technology behind the hype. At much the same time that ultrasonics was being developed as a potential world-beater, other "new" technologies were coming to the fore, in a few cases leading to new vocabulary. One of the new words, tribology, has become a mainstream mechanical engineering discipline, and the IMechE collectively and prominent individual members of the Institution can take a lot of the credit for its development. A second, terotechnology, has lost currency as a term in its own right, but is now part of a whole subset of engineering and management functions to do with maintenance, reliability and design life; so it is still important, even if many engineers would now be hard pressed to remember the actual word.

Other "new" technologies from the 1960s such as control theory, heat transfer, composite materials and a later addition to the vocabulary, mechatronics, have also proved to have staying power and wide application.

The 1960s were, though, the last era when, as in Victorian times, there was a widespread faith in the inevitability of progress and in the ability of technology to solve mankind's problems. Technology itself was a fashionable concept, and parts of British engineering industry were also able, fleetingly, to take advantage of worldwide phenomenon, more to do with the Beatles than with industry's products, which equated British-ness with sophistication.

There were plenty of contenders for technologies, or areas of research, which would take mankind forward. The whole of space exploration and research, for example, was forecast to have immense benefits; new materials, particularly composites but also alloys of titanium and other expensive metals, were going to revolutionise engineering; and then there were the potential earthbound applications from the electronics and computing used in space research, and the potential for manufacturing anything from pharmaceuticals to microchips in the ultra-clean surroundings of a lunar landscape. In some cases, the ideas now sound far-fetched. In others, though, it is simply that the timescale imagined 30 years ago now looks outrageously optimistic.

There was optimism elsewhere. In parallel with these promising fields for future development, there seemed also to be in some areas a view of the innovation process itself which now seems strangely naive: a feeling that a demand only had to be articulated for technology somehow to be able to provide the means for its satisfaction.

Within reason, of course, there is a degree to which technology can be ordered "à la carte". The internal combustion engine lobby at the time of the earliest motor cars saw its apparent advantages as likely to disappear unless the problems of carburation could be overcome: they duly were. A decade earlier, many eminent engineers including the steam turbine pioneer Charles Parsons were convinced it was only a matter of time before developments in engines and in engineering science allowed powered flight: again, they were right, though not many of them were looking towards North Carolina when the Wright Brothers achieved the first aircraft flight in 1903.

It is one thing to anticipate technology developments when all the ingredients are there and when targets and forecasts are firmly based in reality: it is another thing entirely to make predictions of technological developments based on science or engineering which does not yet exist. In the 1960s, the decade when it seemed that technology had no limits, some of the forecasts for mankind's progress were more expressions of faith than realistic targets.

Perhaps the clearest expression of the faith in the inevitability of technological progress was a list compiled in 1967 by eminent

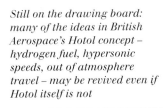

Still on the drawing board: many of the ideas in British Aerospace's Hotol concept – hydrogen fuel, hypersonic speeds, out of atmosphere travel – may be revived even if Hotol itself is not

American technical experts of 100 innovations "very likely in the last third of the 20th century". The list read like a "wish-list" rather than a studied forecast rooted in existing technology. It included controlled hibernation for humans, interplanetary travel, individual flying platforms, programmable dreams and artificial moons for illuminating large areas at night.

None of these have happened, and in fact fewer than a quarter of the 100 ideas have come about, even on a fairly generous interpretation of what were often vague suggestions. The successes include the space shuttle, advances in genetic engineering, all-pervasive computers at work and at home, and direct broadcasting by satellite. Rather more than a quarter of the total are still at least as far away now as they were in 1967, perhaps further, as the chances of serious research being done in some areas has receded in the past 30 years.

Technology prediction of this sort seems to owe as much to science fiction as to engineering fact and has very much fallen out of favour now. In the 1950s and 1960s, though, the mechanisms of innovation were less clearly understood, and there was consequently much more faith both in the ability of science and technology to come up with the innovations that were wanted. There appeared also to be a close cause-and-effect relation between "blue skies" research and new products. These days there is more understanding of the processes, and present attempts to enhance Britain's competitiveness through technology forecasting appear to start from a realistic appreciation of where today's trends are leading, rather than relying on wish-lists.

Military muscle: the artists impression of the Future Offensive Aircraft for British Aerospace may appear unrealistically futuristic, but concepts such as flying wing mean that it is rooted firmly in fact

Of the fundamental all-embracing technologies or areas of research promise which were being developed in the 1960s, electronics and computing seem to have fulfilled more than most of the potential that was forecast for them. One reason for this was that the development of the microprocessor, based on one of the commonest materials on earth, allowed dramatic reductions in cost. And cost, rather more than technology, has been the main driving force in engineering innovation for most of the last third of the 20th century.

Engineers in Britain may have been more used than their counterparts in some other countries to the need to reconcile the drive for innovation with cost control. Successive economic crises since the Second World War created a long and not-very-honourable tradition of aborting technology-led projects – when British Rail's Advanced Passenger Train headed for the siding, it would have found, on the airfield next door, the independent British rocket programme and the TSR-2 aircraft. As a public sector project in the 1970s, the Channel Tunnel came into the same category.

For many British engineers, though, it was an event from the private sector of industry rather than the high politics of public sector research and development that brought home the need for engineer-

Back to earth: the Rolls-Royce RB211 family and its continuing development 30 years on indicate that successful innovation is achieved over the long run, not the short term

ing realism. The Rolls-Royce RB-211 series of aero-engines was, and is, a technological triumph, and one of the British engineering projects of the past 150 years of which it could truly be said that innovations were sent spinning off in all directions. As a feat of engineering, it was glorious. But financially, it was a disaster, and its runaway costs ruined – temporarily – the proudest name in British engineering.

The Rolls-Royce collapse was followed in Britain by other high profile engineering projects, such as power stations, which ran badly over budget and reinforced the need for economy.

But there have been other factors operating worldwide that have had the effect of reducing technology ambition from the peaks of the 1960s. The oil price shocks of the early and late 1970s, the rise of industrial competition from low-cost countries, welfare pressures – all were factors in turning industry and engineering worldwide away from extravagance and into a new era of economy. And in many areas, technology itself was seen as part of the extravagance. The reduction of ambition worldwide from the 1960s to the economic realism of the later years was exemplified by the way the twin technological emblems of the 1960s, the space programmes of both the USA and the Soviet Union, were scaled down from the high adventure of the race to the moon to a more utilitarian practicality of space stations and reusable shuttles.

But there was a still wider phenomenon which has affected technology worldwide. The faith in technology as a driving force for mankind's progress was threatened by familiarity and by better appreciation of the capacity of technology for creating instruments of doubtful benefit, as well as obvious merit. Since man landed on the moon, mankind has become much less starry-eyed about technology.

Television, itself a product of engineering, brought the images of the moon landing back, but also put destruction and warfare on to TV screens in homes worldwide; accidents and disasters, sometimes much exaggerated, generated awareness of some of the less palat-

Learning from the past I: when the Hindenburg airship plunged to earth at Lakehurst, New Jersey in 1937, a whole strand of engineering development was halted. Only in recent years has the airship concept been revived

Learning from the past II: poor engineering design, construction and operation were contributory factors in the Chernobyl disaster in the Ukraine in 1986

able side-effects of technology and industrialisation; the undeniable environmental impact of technology and development was seen as a mixed benefit and the use of rare resources as unsustainable.

Large sectors of engineering technology have become controversial and unpopular in the past 25 years, sometimes with good reason, but sometimes on the flimsiest evidence. The change has affected the work that many professional engineers now do. In many parts of mechanical engineering, rather than contributing to a remorseless forward-looking process of innovation, engineers are going back to the designs of yesteryear, to see whether engineering products and machinery cannot be made better or more efficiently, or perhaps using fewer natural resources such as fuels and natural materials.

The IMechE has an important part to play in this mechanical engi-

Opposite: Learning from the past III: the dangers of the search for the basic commodities of everyday living were highlighted in the Piper Alpha tragedy

The old challenge: engineering's first challenge is overcoming nature. Earthquakes such as the Kobe disaster of 1995 demonstrate that nature is a fickle and powerful adversary

neering voyage of rediscovery, as it has also in helping engineers learn from the mistakes of the past through its conferences and publications on environmental themes and on accidents and disasters.

In this context, the notion of progress through mechanical engineering is a lot more complex than it was for the mechanical engineers of 1847, or even for those of 1947. Progress through engineering is no longer a straightforward process of invention, production, consumption and inevitable benefit, if it ever was, and the modern-day IMechE has to take account of a much wider range of demands on its members than faced by the Victorians who founded the Institution.

Continuing challenges: the threat of dwindling natural resources is a continuing theme of engineering in the late 20th century

There is a much wider range of motivations for people wanting to become mechanical engineers as well. The new pressures on IMechE members are bringing new kinds of people into the profession: people whose aim is to rectify through engineering the mistakes of the past and who intend to create machines and products that will create wealth without destroying the earth's riches.

There are other complexities in modern-day engineering compared with that of 1847. Many of the distinctions between different branches of the engineering profession – some of which have occurred only in the past 50 years – are now becoming blurred through the controlling technologies of electronics and computing. A mechanical engineer may very well, these days, sit in front of a computer screen and talk to the world through a mouse, a modem and an e-mailbox. Mechanical engineering products such as cars are developing through the use of technologies which are not mechanical in origin or in execution. This might well be, to the founders of the IMechE in 1847, the most confusing aspect of present-day mechanical engineering and they would also have great difficulty coming to terms with the greatly accelerated product development that the new technologies allow.

But if these factors might seem to some people to threaten the future of mechanical engineering as a separate discipline with an illustrious future as well as an illustrious past, then there are other indications which suggest exactly the opposite – a wider role with greater influence.

New challenges I: nuclear power and the problems of nuclear waste have been one of the contentious engineering issues of recent years. This is the Sellafield complex in Cumbria

The present-day Institution of Mechanical Engineers is very much concerned with expanding the range and remit of engineers and engineering, as were the pioneers of 150 years ago in 1847, whose work was concentrated largely in the areas of transport, manufacturing and energy.

"New" areas of influence today include medicine: the IMechE has had a group, now a division, since the mid 1960s working with surgeons, physicians and also with other engineering disciplines on the "ultimate" mechanism, the human body. Their work brings together mechanics, tribology, control technology and other parts of engineering. Engineers can never replace doctors, but they can bring skills to surgery and medicine that no doctor's training could ever provide.

Other parts of engineering throw up new challenges for mechanical engineers too. The possibility of electronic engineers working on atomic and subatomic particles as substrates for new generation nanotechnological devices probably presupposes that the mechanical

253

Opposite: New challenges II: the next phases of North Sea development involve extracting oil from deeper and deeper waters. This is one solution, Conoco's tension leg platform for the NW Hutton field

engineers will have had to get there first – to create the nanotechnological machinery to handle, shape, assemble or isolate the minute electronic devices.

Mechanical engineers face challenges that are, literally, of global importance. Measures to protect the environment will depend on contributions from engineers at a local level to reduce waste and develop alternative technologies for existing products and processes.

All visions of the future, from science fiction to science fact, from realistic views of the early years of the next millennium to fanciful fantasies of the millennium after that, rely on the products of

Mechanical engineering experience: an impressive gathering of past IMechE presidents in 1995. Front row: George Adler (1981), Brian Kent (1994), Sir Robert Lickley (1971), Sir Diarmuid Downs (1978); second row: Waheeb Rizk (1984), John Osola (1982), Michael Neale (1990), Cecil French (1988), Paul Fletcher (1975); third row: Gordon Dawson (1979), Oscar Roith (1987), Richard Pike (director-general since 1993), Tony Denton (1993); back row: Tom Patten (1991), Sir Bernard Crossland (1986), Alex McKay (secretary, 1976-1987), Bryan Hildrew (1980)

mechanical engineering and the continued innovation of mechanical engineers. Maybe, one day, all 100 of the innovations "very likely in the last third of the 20th century" will come to fruition: if so, most of them will require mechanical engineering skills in design, manufacture, measurement, control – in innovation.

In 1847, like-minded engineers decided that, by pooling their knowledge and by learning from each other through the organisation of the Institution of Mechanical Engineers, they might make the products and the processes of engineering more widely and more quickly available. They had no doubts that the progress that would come from their work would enrich their own existences and benefit the generations that would follow.

A matter of 150 years on, an age that would see itself as more sophisticated may regard the IMechE's founders as idealistic in their faith in mankind's progress and may envy the Victorian engineers for

Progress into the future...

their certainties and for the relative simplicity of their world. But even hardened cynics can marvel at the way that the Institution that the Victorian engineers created has adapted itself to the technical, industrial and political complexities of the modern world and can claim now more relevance to more people than ever before.

George Stephenson and his mechanical engineering friends in Birmingham in January 1847 started something that, 150 years on, would make them very proud.

Appendixes

IMechE Founder Members, Presidents and Secretaries

1

IMechE Founder Members

Name and dates (where known)	Position or address as given in register book in 1847	Notes
1 **Alexander Allan** – ?1809-1891	Locomotive Department, London & North Western Railway, Crewe	Left Crewe in 1853. Locomotive Superintendent of the Scottish Central Railway to 1865
2 **Charles F Beyer** – 1813-1876	Sharp Bros, Atlas Works, Manchester	German emigré. Formed Beyer, Peacock at Gorton Foundry with Richard Peacock in 1854
3 **Henry Birley**	Engineer, Didsbury, Manchester	The Birleys were an important Manchester engineering dynasty in the first half of the 19th century, and helped James Nasmyth set up at Patricroft. Member until 1892
4 **George Hinton Bovill** – 1821-1868	Engineer, Millwall, Poplar, London	Later partner, Swayne & Bovill, 19 Abchurch Lane, London. Patents on atmospheric railways, gas transmission and manufacturing flour from wheat
5 **James Brown**	James Watt & Co, Soho Works, Birmingham	Resigned in 1875
6 **William Buckle** – 1794-1863	James Watt & Co, Soho Works, Birmingham	Manager of Boulton's mint for 30 years until 1851, when he was appointed superintendent of the Royal Mint in London. Resigned in 1856
7 **Henry Chapman**	2 Upper Charles Street, Westminster	No longer a member in 1855
8 **John Edward Clift** – 1817-1875	Engineer, Birmingham & Staffordshire Gas Works, Bordesley, Birmingham	
9 **Edward Alfred Cowper** – 1819-1893	Fox, Henderson & Co, London Works, Birmingham	Later a consultant at 35A Great George Street, Westminster. IMechE president in 1880-81
10 **Benjamin Cubitt** – 1795-1848	Great Northern Railway, Lincoln	Earlier locomotive superintendent, South Eastern Railway. His family was prominent in construction and civil engineering
11 **Henry Dübs** – 1816-1876	Tayleur's Vulcan Foundry, Warrington	Like Beyer, a German emigré. Later worked for Neilson & Co in Glasgow, then founded Dubs & Co at Polmadie
12 **Edward Fletcher** – 1807-1889	Locomotive Superintendent, York, Newcastle & Berwick Railway, Gateshead	Locomotive superintendent of the North Eastern Railway from 1854 to 1883
13 **Benjamin Fothergill** – 1802-1874	Roberts, Fothergill & Dobinson, Globe Works, Manchester	Manager of Sharp, Roberts & Co to 1845. In partnership with Richard Roberts and Robert Dobinson 1845 to 1850. Laid out textile machinery section at Great Exhibition of 1851. Died in poverty. 'Erased' from membership, 1870.
14 **Charles Fox** – 1810-1874	Fox, Henderson & Co, 8 New Street, Spring Gardens, London	Prepared the detailed design of the Crystal Palace. Knighted. Resigned, 1871
15 **William S Garland** – ?1814-?	James Watt & Co, Soho Works, Birmingham	Resigned, 1873
16 **Charles Geach** – 1808-1854	Birmingham & Midland Bank, Union Street, Birmingham	Later a Member of Parliament
17 **Benjamin Goodfellow** – 1811-1863	Hyde Iron Works, Hyde, Manchester	Patents on steam engines
18 **John Henderson** – ?-1858	Fox, Henderson & Co, London Works, Birmingham	Partner of Charles Fox

Name and dates (where known)	Position or address as given in register book in 1847	Notes
19 **George Hennet** – 1799-1857	Engineer, 24 Duke Street, Westminster	Atmospheric railways
20 **John Hick** – 1815-1894	Soho Iron Works, Bolton	No longer a member in 1855, but re-elected in 1871. Son of Soho Iron Works founder, formed Hick Hargreaves, making textile machinery and steam engines, mainly stationary
21 **James Ward Hoby**	Engineer, 6 Clifton Terrace, Brighton	Later at London Works, Renfrew, near Glasgow. Resigned 1856
22 **Paul Rapsey Hodge**	140 Strand, London	Later at 11 Buckingham Street, Adelphi, London. Resigned 1856
23 **John Horton jr**	Parkfield Iron Works, Wolverhampton	Resigned 1850
24 **Joseph Howell**	Hawarden Iron Works, Flintshire	
25 **Edward Humphrys** – 1808-1867	Dockyard, Woolwich	Formerly with G & J Rennie. Later with Humphrys, Tennant & Dykes, Deptford Pier, London. Resigned 1856
26 **W Hutchinson**	Robert Stephenson Locomotive Works, Newcastle upon Tyne	Resigned 1851
27 **Peter Rothwell Jackson** – 1813-1899	Salford Rolling Mills, Manchester	Patents on wheel manufacture, hydraulic presses and pumps, textile machinery.
28 **Edward Jones**	Bridgewater Iron Works	Address given as 3 Greenfield Terrace, Edge Hill, Liverpool in 1855. Resigned 1874
29 **Matthew Kirtley** – 1813-1873	Locomotive Superintendent, Midland Railway, Derby	Patent for railway wheels
30 **William Langdon**	James Watt & Co, Fenchurch Street, London	Resigned, 1850
31 **Edmund Leahy**	32 Charing Cross, London	Resigned, 1848. Patent for locomotive carriage
32 **Matthew Leahy**	12 Rue de Chateau, Boulogne, France	Resigned, 1848. Patent for steam engine
33 **Sampson Lloyd** – 1808-1874	Lloyds Fosters & Co, Old Park Iron Works, Wednesbury, Staffordshire	Iron tyre patent
34 **William Lee**	Sharebroker, Cheltenham	Resigned, 1849
35 **James Edward McConnell** – 1815-1883	Locomotive Superintendent, London & North Western Railway, Wolverton	Resigned, 1861
36 **William Middleton** – 1795-1868	Vulcan Iron Foundry, Summer Lane, Birmingham	
37 **Joseph Miller** – 1797-1860	Miller & Ravenhill, Marine Engineers, Blackwall, London	Living in 1855 at Oakley House, Alpha Road, St John's Wood, London. Fellow of the Royal Society
38 **John Milner**	Pembroke Dock, Pembroke	Resigned, 1848
39 **William Owen** – 1810-1881	Sandford & Owen, Phoenix Forge, Rotherham	
40 **Frederick Parker**	Engineer, Great Northern Railway, Louth, Lincolnshire	Resigned, 1851
41 **Richard Peacock** – 1820-1889	Locomotive Superintendent, Manchester, Sheffield & Lincolnshire Railway, Manchester	From 1854 partner in Beyer, Peacock & Co, Gorton, Manchester. Later Member of Parliament
42 **Arthur Potts** – 1814-1888	Jones & Potts, Viaduct Foundry, Newton-le-Willows, Lancashire	Resigned, 1851
43 **Joseph Radford**	St George's Foundry, Minshull Street, Manchester	Resigned, 1848. Patent for envelope making machinery
44 **John Ramsbottom** – 1814-1897	Locomotive Superintendent, London & North Western Railway (Manchester & Birmingham Section), Manchester	Locomotive Superintendent, London & North Western Railway, Crewe, from 1857 to 1871, retiring through ill-health. Later consultant to Lancashire & Yorkshire Railway, Horwich. IMechE President 1870-71.

Name and dates (where known)	Position or address as given in register book in 1847	Notes
45 **Thomas Richards**	Engineer, Gas Works, Worcester	Living in 1855 at 30 Hill Morton Road, Rugby. Erased, 1858
46 **Henry Robinson**	Charles Robinson & Son, Eaton Lane, Pimlico, London	Address in 1855 is 8 Chandos Street, Cavendish Square, London. Resigned, 1857
47 **David M Roche** – ?-1857	Locomotive Superintendent, London & North Western Railway (Huddersfield & Manchester Section), Huddersfield	Working in 1855 on the East Indian Railway, Howrah, Calcutta.
48 **Archibald Slate** – 1815-1860	Woodside Iron Works, Dudley	Resigned, 1856
49 **Henry Smith**	Vulcan Iron Works, West Bromwich	Register says he resigned in 1850. Another Henry Smith, from Liverpool, had joined in 1848. One of them was working at the Victoria Iron Works, Smethwick, in 1855
50 **George Stephenson** – 1781-1848	Tapton House, Chesterfield	
51 **Robert Stephenson** – 1803-1859	24 Great George Street, Westminster	IMechE president 1849-53; life member
52 **Edward Tayleur**	Tayleur & Co, Vulcan Foundry, Warrington	A member of the family that employed Henry Dübs. Resigned, 1854.
53 **Thomas Walker** – ?1816-1887	Patent Shaft Works, Wednesbury, Staffordshire	Resigned, 1872. Many steam engine patents in his name
54 **William Weallens** – 1823-1862	Robert Stephenson & Co, Locomotive Works, Newcastle upon Tyne	
55 **Richard Williams** – 1817-1909	Patent Shaft Works, Wednesbury, Staffordshire	The longest-surviving founder member, later made a life member
56 **Joseph Woods** – 1816-1849	Engineer, Barge Yard Chambers, Bucklersbury, London	Patents on rotary engines, power transmission, copying machines and springs

2

IMechE Presidents

No	Presidential years	Name	Dates	Sphere of influence
1	1847-48	**George Stephenson**	1781-1848	Railway pioneer.
2	1849-53	**Robert Stephenson**	1803-1859	Railway and civil engineer, MP
3	1854-55	**William Fairbairn**	1789-1874	Manufacturer, trader, ironmaster, bridges, mill wheels, ships. Later created baronet
4	1856-57 1866	**Joseph Whitworth**	1803-1887	Screw threads, machine tools, precision measurement, armaments. Later created baronet. Major benefactor of engineering education
5	1858-59 1867-68	**John Penn**	1805-1875	Marine steam engines. Invented lignum vitae bearing for propeller shafts
6	1860	**James Kennedy**	1797-1886	Marine engines and locomotives
7	1861-62 1869	**William George Armstrong**	1810-1900	Trained as lawyer. Hydraulic machinery and armaments. First to install domestic electric light. Knighted, then made a peer.
8	1863-65	**Robert Napier**	1791-1876	Shipbuilding and marine engines on Clydeside.
9	1870-71	**John Ramsbottom**	1814-1897	Locomotive superintendent, London & North Western Railway. Inventor of piston rings, and of water troughs for railway locomotives
10	1872-73	**Charles William Siemens**	1823-1883	German-born metallurgist and electrical engineering pioneer. Later knighted
11	1874-75	**Sir Frederick Joseph Bramwell**	1818-1903	Steam engines and boilers
12	1876-77	**Thomas Hawksley**	1807-1893	Water and gas supply engineer
13	1878-79	**John Robinson**	1823-1902	Steam engines
14	1880-81	**Edward Alfred Cowper**	1819-1893	Wrought iron structures, including Crystal Palace roof. Regenerative heating for blastfurnaces
15	1882-83	**Percy Graham Buchanan Westmacott**	1830-1917	Hydraulic machinery and armaments. Colleague of Armstrong.
16	1884	**Sir Isaac Lowthian Bell**	1816-1904	Iron and steel making
17	1885-86	**Jeremiah Head**	1835-1899	Steam-powered agricultural machinery, rolling mills
18	1887-88	**Sir Edward Hamer Carbutt**	1837-1905	Iron and steel making
19	1889	**Charles Cochrane**	1835-1898	Iron and steel making
20	1890-91	**Joseph Tomlinson**	1823-1894	Locomotive superintendent, Taff Vale Railway and Metropolitan Railway
21	1892-93	**Sir William Anderson**	1835-1899	Builder of bridges and factories worldwide
22	1894-95	**Professor Alexander Blackie William Kennedy**	1847-1928	Professor of engineering at University College, London. Electric power generation and transmission. Later knighted
23	1896-97	**Edward Windsor Richards**	1831-1921	Iron and steel making
24	1898	**Samuel Waite Johnson**	1831-1912	Chief mechanical engineer, Great Eastern Railway and Midland Railway

No	Presidential years	Name	Dates	Sphere of influence
25	1899-1900	**Sir William Henry White**	1845-1913	Director of naval construction for the Royal Navy and for navies worldwide
26	1901-02	**William Henry Maw**	1838-1924	Editor, Engineering
27	1903-04	**Joseph Hartley Wicksteed**	1842-1919	Testing machines and machine tools
28	1905-06	**Edward Pritchard Martin**	1844-1910	Iron and steel making
29	1907-08	**Tom Hurry Riches**	1846-1911	Chief engineer, Taff Vale Railway
30	1909-10	**John Audley Frederick Aspinall**	1851-1937	Chief mechanical engineer, Lancashire & Yorkshire Railway. Later knighted
31	1911-12	**Edward Bayzard Ellington**	1845-1914	Hydraulic machinery design and manufacture.
32	1913-14	**Sir Hay Frederick Donaldson**	1856-1916	Chief engineer, Royal Ordnance Factories, and chief engineering adviser, Ministry of Munitions
33	1915-16	**William Cawthorne Unwin**	1838-1933	Oil engine research. Engineering professor
34	1917-18	**Michael Longridge**	1847-1928	Chief engineer, British Engine insurance group. Family connected with George Stephenson through Bedlington Iron Works.
35	1919	**Edward Hopkinson**	1859-1922	Electric traction. Managing director, Mather & Platt. Member of Parliament.
36	1920-21	**Captain Matthew Henry Phineas Riall Sankey**	1853-1925	Military engineer, then oil engines and wireless telegraphy
37	1922	**Dr Henry Selby Hele-Shaw**	1854-1941	First professor of mechanical engineering, Liverpool University
38	1923	**Sir John Dewrance**	1858-1937	Inventor. Chairman, Babcock & Wilcox. Married Richard Trevithick's grand-daughter
39	1924	**William Henry Patchell**	1862-1932	Electricity supply in mining and other industries
40	1925	**Sir Vincent Litchfield Raven**	1858-1934	Chief mechanical engineer, North Eastern Railway
41	1926	**Sir William Reavell**	1866-1948	Founder, air compressor manufacturing company
42	1927	**Sir Henry Fowler**	1870-1938	Chief mechanical engineer, Midland Railway and London Midland & Scottish Railway
43	1928	**Richard William Allen**	1867-1955	Pumps and marine equipment. Chairman, W H Allen.
44	1929	**Daniel Adamson**	1869-1930	Gears, cranes, cutting tools
45	1930	**Loughnan St Lawrence Pendred**	1870-1953	Editor, The Engineer, 1906-46
46	1931	**Edwin Kitson Clark**	1866-1943	Locomotive engineer
47	1932	**William Taylor**	1865-1937	Lens manufacturing
48	1933	**Alan Ernest Leofric Chorlton**	1874-1946	Pumps and diesel engines. Member of Parliament
49	1934	**Charles Day**	1867-1949	Steam and diesel engineer
50	1935	**Major-General Alexander Elliott Davidson**	1880-1962	Mechanised military transport
51	1936	**Sir Herbert Nigel Gresley**	1876-1941	Chief mechanical engineer, Great Northern Railway and London & North Eastern Railway
52	1937	**Sir John Edward Thornycroft**	1872-1960	Shipbuilding, motor vehicle design and manufacture
53	1938	**David E Roberts**	1867-1950	Iron and steel manufacture
54	1939	**E Bruce Ball**	1873-1944	Motor vehicles, hydraulic valves
55	1940	**Asa Binns**	1873-1946	Engineer, Port of London

No	Presidential years	Name	Dates	Sphere of influence
56	1941	**Sir William Arthur Stanier**	1876-1965	Chief mechanical engineer, London Midland & Scottish Railway
57	1942	**Colonel Stephen Joseph Thompson**	1875-1955	Boilers
58	1943	**Frederick Charles Lea**	1871-1952	Engineering professor at Birmingham and Sheffield. Aircraft and engine research
59	1944	**Harry Ralph Ricardo**	1885-1974	Internal combustion engine research and design
60	1945	**Andrew Robertson**	1883-1970	Professor of mechanical engineering, Bristol University
61	1946	**Oliver Vaughan Snell Bulleid**	1882-1970	Chief mechanical engineer, Southern Railway
62	1947	**Lord Dudley Gordon**	1883-1972	Refrigeration engineering. Chairman, J & E Hall
63	1948	**E William Gregson**	1891-1977	Marine engines
64	1949	**Herbert John Gough**	1890-1965	Engineering research at the National Physical Laboratory
65	1950	**Stanley Fabes Dorey**	1891-1972	Chief engineer surveyor, Lloyd's Register of Shipping
66	1951	**Arthur Clifford Hartley**	1889-1960	Chief engineer, Anglo-Iranian Oil Co. Wartime inventor of Pluto and Fido
67	1952	**Sir David Randall Pye**	1886-1960	Air Ministry research engineer
68	1953	**Alfred Roebuck**	1889-1962	Engineering metallurgy
69	1954	**Richard William Bailey**	1885-1957	High temperature steel and materials research
70	1955	**Percy Lewis Jones**	1886-1966	Marine engines and shipbuilding
71	1956	**Thomas Arkle Crowe**	1896-1972	Marine engines. Engineering director, John Brown
72	1957	**Sir George Nelson**	1887-1962	Chairman, English Electric. Later created 1st Lord Nelson of Stafford
73	1958	**Air Marshal Sir Robert Owen Jones**	1901-1972	Aircraft engineer
74	1959	**Herbert Desmond Carter**	1906-1990	Diesel engines. Chairman, Crossley Bros
75	1960	**Owen Alfred Saunders**	1904-1993	Professor of mechanical engineering, Imperial College, and vice-chancellor, London University. Later knighted
76	1961	**Sir Charles Kenneth Felix Hague**	1901-1974	Chairman, Babcock & Wilcox
77	1962	**John Hereward Pitchford**	1904-1995	Internal combustion engines. Chairman, Ricardo
78	1963	**Roland Curling Bond**	1903-1980	Chief mechanical engineer, British Railways
79	1964	**Vice-Admiral Sir Frank Trowbridge Mason**	1900-1988	Engineer-in-Chief, Royal Navy
80	1965	**Harold Norman Gwynne Allen**	1912-1995	Power transmission, W H Allen
81	1966	**Lord (Christopher) Hinton of Bankside**	1901-1983	Initially railway engineer, then power generation, including early days of nuclear power
82	1967	**Hugh Graham Conway**	1914-1989	Aero-engines and gas turbines with Rolls-Royce
83	1968	**Arnold Lewis George Lindley**	1902-1995	Chairman of GEC, then Motherwell Bridge. Knighted
84	1969	**Donald Frederick Galloway**	1913-1997	Manufacturing and machine tool engineer. Production Engineering Research Association.
85	1970	**John Lamb Murray Morrison**	1906-	Professor of mechanical engineering, Bristol University
86	1971	**Robert Lang Lickley**	1912-	Aircraft engineer. Managing director, Hawker Siddeley Aviation
87	1972	**Donald Gresham Stokes**	1914-	Automobile engineer. Chairman, managing director and chief executive, British Leyland. Life peer

No	Presidential years	Name	Dates	Sphere of influence
88	1973	John William Atwell	1911-	Steel industry and pump manufacture with Weir Group. Knighted
89	1974	St John de Holt Elstub	1915-1989	Metals industry with IMI. Knighted
90	1975	Paul Thomas Fletcher	1912-	Process plant and nuclear power plant. Influential in setting up technicians' institutions
91	1976	Ewen McEwen	1916-1993	Chief engineer, Lucas
92	1977	Hugh Ford	1913-	Professor of mechanical engineering, Imperial College, London. Knighted.
93	1978	Diarmuid Downs	1922-	Internal combustion engine research, Ricardo. Knighted.
94	1979	James Gordon Dawson	1916-	Chief engineer, Shell Research. Technical director, Perkins Engines. Managing director, Zenith Carburettors
95	1980	Bryan Hildrew	1920-	Managing director, Lloyd's Register of Shipping
96	1981	Francis David Penny	1918-	Director, National Engineering Laboratory, Yard consultancy.
97	1982	Victor John Osola (Väinö Juhani Osola)	1926-	Process engineer specialising in safety glass
98	1983	George Fritz Werner Adler	1926-	Research director, British Hydromechanical Research Association
99	1984	Waheeb Rizk	1921-	Gas turbines at GEC
100	1985	Sir Philip Foreman	1923-	Aerospace engineer. Chief executive, Short Brothers
101	1986	Bernard Crossland	1923-	Professor of mechanical engineering, Queen's University Belfast. Knighted in 1990
102	1987	Oscar Roith	1927-	Chief engineer and scientist, Department of Trade and Industry
103	1988	Cecil Charles John French	1926-	Internal combustion engines. Director, Ricardo.
104	1989	Roy Ernest James Roberts	1928-1993	Director, GKN and Simon Engineering
105	1990	Michael Cooper Neale	1929-	Gas turbines, defence procurement.
106	1991	Duncan Dowson	1928-	Professor of engineering fluid mechanics and tribology, Leeds University.
107	1992	Tom D Patten	1926-	Offshore engineering. Professor of mechanical engineering, Heriot-Watt University. Chairman, Marine Technology Directorate
108	1993	Anthony Albert Denton	1937-	Offshore engineering. Chairman, Noble Denton Group
109	1994	Brian Hamilton Kent	1931-	Design and engineering management. Chairman, Wellington Holdings
110	1995	Frank Christopher Price	1946-	Technical director, Rolls-Royce Industrial and Marine Power
111	1996	Robert William Ernest Shannon	1937-	Inspection engineering with gas industry. Executive director, ERA Technology
112	1997	Pamela Liversidge	1949	Powder metallurgy. Managing director, Quest Investments

3

IMechE Secretaries

No	Years in office	Name	Dates	Notes
1	1847-1848	**Archibald Slate**	1815-1860	First honorary secretary and one of the engineers involved from the beginning in the IMechE. Gave up secretaryship when volume of work dictated that a permanent secretary was needed
2	1848-1849	**Archibald Kintrea**		Took over from Slate and a Mr Maher who was 'acting secretary'. Lasted less than a year. Nothing known of him beyond his name
3	1849-1877	**William Prime Marshall**	?1816-1906	Assistant engineer to Robert Stephenson, then chief engineer of the Norfolk Railway. Retired 'in his 62nd year' when the Institution moved from Birmingham to London. Also held engineering professorship at forerunner to Birmingham University
4	1878-1884	**Walter Raleigh Browne**	1842-1884	Became secretary after move to London was completed. Resigned in January 1884 when by-law was passed requiring secretary to give up outside work. In August 1884, went with the British Association on a visit to Canada: contracted typhoid and died there
5	1884-1897	**Alfred Bache**	1835-1907	Had worked for the IMechE for 29 years, latterly as assistant secretary, before appointment in 1884. Resigned after a long period of ill-health apparently brought on by overwork
6	1897-1920	**Edgar Worthington**	?1859-1934	Chosen from 100 applicants. Managing engineer of Beyer Peacock's Gorton works. Resigned, like Bache, after a long period of ill-health from overwork
7	1920-1937	**Magnus Mowat**	1875-1953	Royal Engineers' brigadier-general who served an engineering apprenticeship on the North British Railway and had been an assistant engineer on the ill-conceived Great Central Railway extension to London. The third secretary in succession to succumb to ill-health, though he recovered sufficiently to help in an honorary capacity through the Second World War
8	1938-1942	**James Edward Montgomrey**	? - 1965	Even IMechE publications sometimes spelled his name wrong. Appointed in 1921 as assistant secretary for technical matters, examinations and membership – one of two assistants intended, in vain, to stop Mowat succumbing to the pressures, as his two predecessors had done
9	1942-1951	**Henry Lewis Guy**	1887-1956	Known (and widely feared) as The Doctor. Fellow of the Royal Society, influential in scientific, research and technical fields before and during the Second World War. Knighted. See Chapter Five
10	1951-1961	**Brian G Robbins**	? - 1976	Previously secretary of the Institution of Automobile Engineers, which merged with the IMechE in 1947
11	1961-1976	**Kenneth Harry Platt**	1909-1985	Assistant secretary to Robbins. Fleetingly an engineering professor at the University of Benares in 1939. Also personnel director at Brush Engines
12	1976-1987	**Alexander Matthew McKay**	1921-	Major-general who joined the IMechE after retiring as director of electrical and mechanical engineering for the army. Stayed on for an extra year to oversee hoped-for merger with Institution of Production Engineers, which was then rejected by members
13	1987-1993	**Ronald William Mellor**	1930-	Former chief manufacturing engineer and director at Ford. Retired after a long period of ill-health
14	1993-	**Richard Andrew Pike**	1950-	The first Director-General and Secretary. Engineer and senior executive with BP, latterly in the Far East

Acknowledgements

Books, like engineering inventions, tend to bear the name of one author, though they are nearly always the result of teamwork. In this instance, the "team" has involved many members of the staff of the Institution of Mechanical Engineers, fellows and members of the IMechE and outside contributors in engineering industry. To all of them, I am grateful for their contributions and their encouragement.

In several cases, this general acknowledgement is insufficient. This book could not have been produced without the knowledge and the commitment of Keith Moore, the IMechE's senior librarian and archivist, who made important contributions to the text and sourced much of the illustrative material. Jacqui Ollerton, manager of the information and library service, was instrumental in maintaining the project's momentum and made many helpful suggestions. Other members of the IMechE staff made significant contributions: Mike Claxton of the library staff is an expert photographer of difficult archive material; his colleague Mike Scanlon read and improved several chapters and all the library staff helped materially in the research; John Dunn, editor of Professional Engineering, was generous with his magazine's photo archives and encouraging in other ways; director of publishing Jamie Cameron co-ordinated the project; and director of finance Robert Howard Jones offered constructive advice and encouragement. I am especially grateful to two former presidents, Tom Patten and Duncan Dowson, who read the text and suggested significant improvements, and to the IMechE director-general Richard Pike for support and commitment.

Some of the original research was done by Hugh Barty-King; Roger Fuller designed the cover; John Duncan and Michael Bush of Dawson Strange Photography were responsible for many of the photographs of archive material; Jo Lee was the book designer; Jeremy Greenwood of Quiller Press oversaw the project from start to finish. Companies and organisations that supplied illustrative material included Rolls-Royce aero-engine and industrial power groups; British Steel/PPL; British Gas; Ford Motor Company; the British Motor Industry Heritage Trust; the Science & Society Picture Library; Hoover; BNFL; British Aerospace; the Press Association; QA Photos and Eurostar. I am grateful to Mr Roderick McConnell for permission to reproduce the painting of his ancestors and to the Lanercost Collection of Birmingham for the line drawing of James McConnell's house.

Inevitably, I will have omitted someone who deserves credit or I will have inadvertently trespassed somewhere across someone's copyright: to them, I apologise. To all the other members of the team: thank you.

John Pullin
June 1997

Index

Italicised page numbers refer to illustrations.